PHOTOGRAPHIC REPRODUCTION

PHOTOGRAPHIC REPRODUCTION

METHODS, TECHNIQUES, AND APPLICATIONS FOR ENGINEERING AND THE GRAPHIC ARTS

Harold Denstman
Supervisor of Photographic Services
Lockheed Electronics Co.

Morton J. Schultz
Technical Communications Consultant

McGRAW-HILL BOOK COMPANY, INC.
New York   Toronto   London

PREFACE

THIS BOOK is a consequence of growth—growth in technology, which in turn has led to growth in industry, which still in turn has led to growth in the services needed by industry.

One would have to search far to find any service, industrial or otherwise, that has increased as rapidly in scope or importance as photographic and engineering reproduction. Practically every office in every plant in the United States and abroad relies on them in one way or another. Some even depend on them for existence.

Yet not until now has there been a book written exclusively to guide the reproduction technician in the performance of his tasks. Naturally, other works devote a portion of their pages to discussions of specific reproduction processes, but these are interspersed in a greater volume of text devoted to such allied fields as general photography and printing.

The purpose of this book is to provide the reproduction technician with a practical working knowledge of the important basic and advanced reproduction processes at his disposal. There are literally hundreds of variations, which manifest themselves in the processes discussed herein. To describe each, however, would require a volume in size many times larger than this one.

The authors have limited the term *reproduction* to those procedures and processes performed by technicians educated, trained, and experienced in the technical aspects of the craft. Certainly, a secretary in a business office performs a reproduction function when she types carbon copies of a document or when she makes copies of that document in an office reproduction machine. In this book, however, the authors are not concerned with those reproduction processes which are easily taught to nontechnical personnel. The title of this work clearly explains its scope.

Photographic reproduction as used in this book refers generally to copying by photographic techniques. For the most part, this type of reproduction employs sensitive materials coated with various types of continuous-tone emulsions and provides as its end product photographic prints, slides, and so forth. The result is a product that does *not* depend for its existence upon the use of a printing press.

An allied, but distinct, branch of photographic reproduction is photomechanical reproduction, which includes all methods used in the preparation of negatives and positives for platemaking, as well as preparation of the plates themselves for the printing press. The final product of photomechanical reproduction is achieved by employing a type of printing press, such as offset printing or silk-screen printing.

Engineering reproduction refers to methods that provide same-size copies of the original by employing various machines, such as diazo and blueprint.

The book is divided into three parts: Techniques of Line Reproduction, Techniques of Continuous-tone Reproduction, and Techniques of Engineering Reproduction. In each is interspersed photographic, photomechanical, and engineering reproduction functions, because all three often overlap one another and it is sometimes necessary to use all three in the achievement of a final product. The book was organized in this manner to provide a more suitable working reference for the reader—a reference that will enable him to pinpoint specific problem areas in which he needs firsthand, immediate information.

The authors wish to thank the many manufacturers and organizations connected with photography and the graphic arts who helped in the preparation of this book. Since these firms are mentioned throughout the text, we have not provided here a long list of those to whom we are grateful.

There is no dedication page in this book. The authors believe that any dedication should go to the thousands of people in the reproduction industry who have made the industry a vital part of American business.

Harold Denstman
Morton J. Schultz

CONTENTS

part I | **TECHNIQUES OF LINE REPRODUCTION**

1 | REPRODUCTION OF LINE MATERIAL IN THE CAMERA

LINE MATERIAL is a line drawing or printed matter consisting of only dark and light areas. It does not possess tonal gradations (intermediate tones) from black to gray to white as does a regular photograph. A page of this text, with its black type on white background, is an example of line material.

When discussing reproduction of line material by means of a camera, one is presented immediately with a problem in semantics. A piece of line material for reproduction by camera is called a line original by reproduction engineers and commercial photographers, line copy by engravers, and a layout or original drawing by draftsmen and artists. To avoid complicacy, this book employs the term *line material* when referring to this area of reproduction. One should be familiar with the terms used by others, however, to ensure a compatibility of understanding when working with those engaged in trades allied to reproduction.

Line material can be presented to the technician for reproduction in one of five forms. It can be a line original, a line duplicate, a manuscript, a piece of previously printed matter, or a composite. When one is asked to reproduce a piece of line material, it is important to determine what form that material takes. This determination enables employment of the correct reproduction technique, issuance of understandable instructions to personnel who will reproduce the material, and establishment of a standardized system for reproducing the various forms.

Each form has distinct characteristics that distinguish it from the others and dictate the method of reproduction. Those seeking reproduction services and those offering such services should use the same terminology

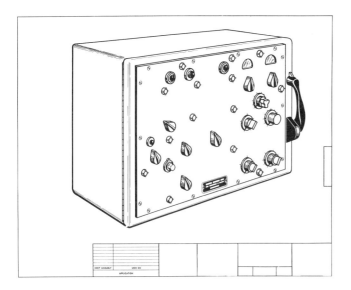

FIG. 1.1   A line original.

when discussing the line material to be reproduced, thus ensuring the desired result.

A *line original* (Fig. 1.1) is an original drawing or layout prepared by a draftsman or artist. It consists usually of pencil or ink lines and symbols, and may have text or lettering callouts included on it to identify particular areas. A line original may be in color as well as in black and white. (Reproduction of a colored line original presents problems different in scope from reproduction of a black-and-white line original. The concepts discussed in Secs. 1.1 and 1.2, as well as methods and techniques explained in other chapters throughout this book, will help overcome these problems.)

A line original can also be in the form of repro copy, which is proof of either hot or cold typeset and text produced by an electric typewriter using a one-time carbon ribbon that creates firm impressions similar to those made by one of the printing processes.

A *line duplicate* is a reproduction of an original piece of line material. This original is not available or usable for further reproduction since it may be torn, faded, or lost. In the absence of the original, the duplicate itself must be reproduced. A line duplicate could have been produced from the original by photographic means or by one of the many reproduction

FIG. 1.2   A line duplicate produced by photostat.

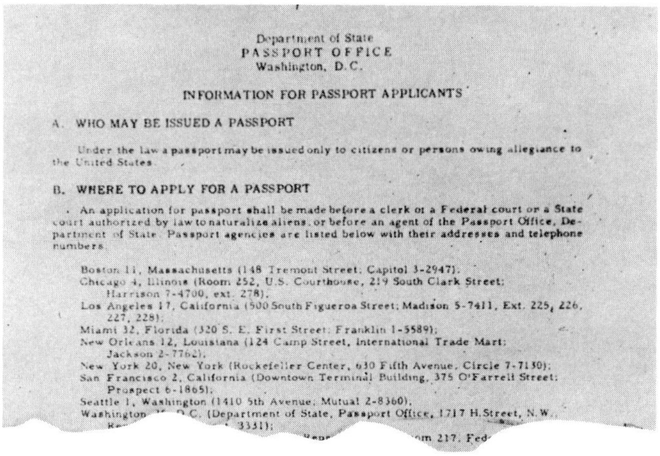

Department of State
PASSPORT OFFICE
Washington, D.C.

INFORMATION FOR PASSPORT APPLICANTS

A. WHO MAY BE ISSUED A PASSPORT

Under the law a passport may be issued only to citizens or persons owing allegiance to the United States.

B. WHERE TO APPLY FOR A PASSPORT

An application for passport shall be made before a clerk of a Federal court or a State court authorized by law to naturalize aliens, or before an agent of the Passport Office, Department of State. Passport agencies are listed below with their addresses and telephone numbers.

Boston 11, Massachusetts (148 Tremont Street; Capitol 3-2947);
Chicago 4, Illinois (Room 252, U.S. Courthouse, 219 South Clark Street; Harrison 7-4700, ext. 278);
Los Angeles 17, California (500 South Figueroa Street; Madison 5-7411, Ext. 225, 226, 227, 228);
Miami 32, Florida (320 S. E. First Street; Franklin 1-5589);
New Orleans 12, Louisiana (124 Camp Street, International Trade Mart; Jackson 2-7762);
New York 20, New York (Rockefeller Center, 630 Fifth Avenue, Circle 7-7130);
San Francisco 2, California (Downtown Terminal Building, 375 O'Farrell Street; Prospect 6-1865);
Seattle 1, Washington (1410 5th Avenue; Mutual 2-8360);
Washington ... D.C. (Department of State, Passport Office, 1717 H Street, N.W., ... 3331);
... om 217, Fed...

FIG. 1.3 A line duplicate produced by thermography.

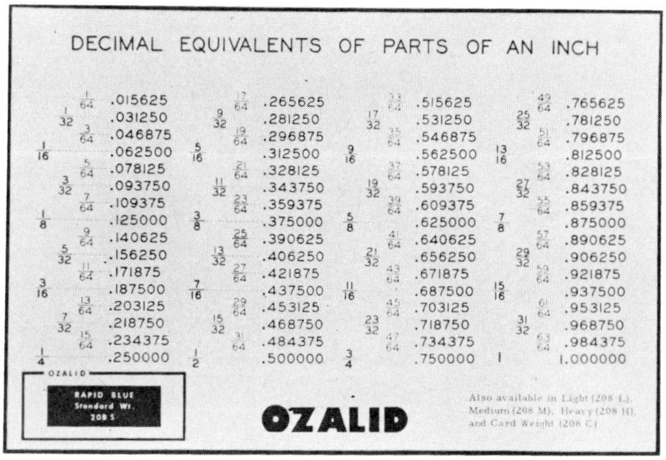

DECIMAL EQUIVALENTS OF PARTS OF AN INCH

1/64	.015625	17/64	.265625	33/64	.515625	49/64	.765625
1/32	.031250	9/32	.281250	17/32	.531250	25/32	.781250
3/64	.046875	19/64	.296875	35/64	.546875	51/64	.796875
1/16	.062500	5/16	.312500	9/16	.562500	13/16	.812500
5/64	.078125	21/64	.328125	37/64	.578125	53/64	.828125
3/32	.093750	11/32	.343750	19/32	.593750	27/32	.843750
7/64	.109375	23/64	.359375	39/64	.609375	55/64	.859375
1/8	.125000	3/8	.375000	5/8	.625000	7/8	.875000
9/64	.140625	25/64	.390625	41/64	.640625	57/64	.890625
5/32	.156250	13/32	.406250	21/32	.656250	29/32	.906250
11/64	.171875	27/64	.421875	43/64	.671875	59/64	.921875
3/16	.187500	7/16	.437500	11/16	.687500	15/16	.937500
13/64	.203125	29/64	.453125	45/64	.703125	61/64	.953125
7/32	.218750	15/32	.468750	23/32	.718750	31/32	.968750
15/64	.234375	31/64	.484375	47/64	.734375	63/64	.984375
1/4	.250000	1/2	.500000	3/4	.750000	1	1.000000

OZALID
RAPID BLUE
Standard Wt.
208 S

OZALID

Also available in Light (208 L), Medium (208 M), Heavy (208 H), and Card Weight (208 C)

FIG. 1.4 A line duplicate produced by diazo.

machines, such as photostat (Fig. 1.2), office copying (Fig. 1.3), diazo (Fig. 1.4), or blueprint. (For a discussion of these machines and the results they produce, see Chap. 7.)

A *manuscript* (Fig. 1.5) is any form of handwriting done with a pencil,

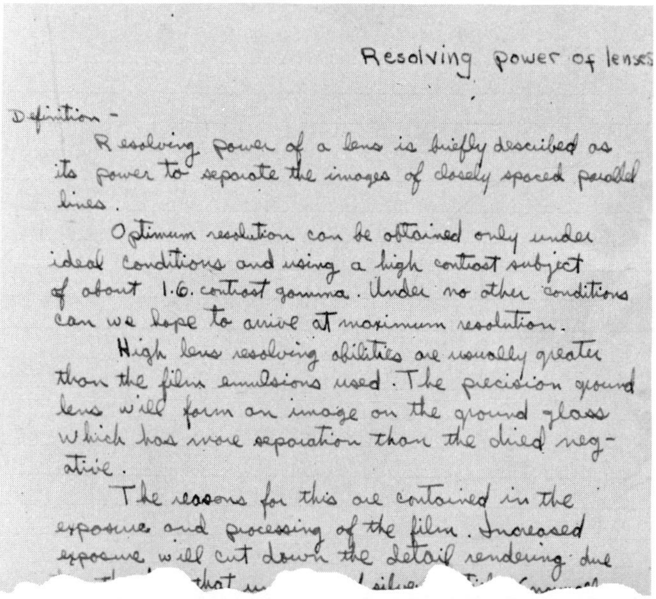

Resolving power of lenses

Definition -

Resolving power of a lens is briefly described as its power to separate the images of closely spaced parallel lines.

Optimum resolution can be obtained only under ideal conditions and using a high contrast subject of about 1.6 contrast gamma. Under no other conditions can we hope to arrive at maximum resolution.

High lens resolving abilities are usually greater than the film emulsions used. The precision ground lens will form an image on the ground glass which has more separation than the dried negative.

The reasons for this are contained in the exposure and processing of the film. Increased exposure will cut down the detail rendering due ... that ... silver ...

FIG. 1.5 Manuscript.

I sera le juge des nations, l'arbitre d'un grand nombre de peuples. De leurs glaives ils forgeront des hoyaux, et de leurs lances des serpes: Une nation ne tirera plus l'epee contre une autre, et l'on n'apprendra plus la guerre.

שָׁפַט בֵּין הַגּוֹיִם וְהוֹכִיחַ לְעַמִּים
רַבִּים וְכִתְּתוּ חַרְבוֹתָם לְאִתִּים
וַחֲנִיתוֹתֵיהֶם לְמַזְמֵרוֹת לֹא־יִשָּׂא
גוֹי אֶל־גּוֹי חֶרֶב וְלֹא־יִלְמְדוּ
עוֹד מִלְחָמָה׃

gli giudichera tra nazione e nazione e sara l'arbitro fra molti popoli; ed essi delle loro spade fabbricheranno vomeri d'aratro, e delle loro lance, roncole; una nazione non levera piu la spada contro un'altra, e non impareranno piu la guerra.

ele exercera o seu juizo sobre as gentes, e reprendera a muitos povos; e estes converterao as suas espadas em enxadoes e as suas lancas em foices: nao levantara espada nacao contra nacao, nem aprenderao mais a guerrear.

juzgará entre las gentes, y reprenderá á muchos pueblos, y volverán sus espadas en rejas de arado, y sus lanzas en hoces: no alzará espada gente contra gente, ni se ensayarán más para la guerra.

ch han skall doma mellan hednafolken och skipa ratt at manga folk. da skola de smida sina svard till plogbillar och sina spjut till vingardsknivar. folken skola ej mer lyfta svard mot varandra och icke mer lara sig att strida

nd er wird richten unter den Heiden und strafen viele Völker. Da werden sie ihre Schwerter zu Pflugscharen und ihre Spiesse zu Sicheln machen. Denn es wird kein Volk wider das andere ein Schwert aufheben, und werden hinfort nicht mehr kriegen lernen.

будет Он судить народы, и обличит многие племена; и перекуют мечи свои на орала, и копья свои — на серпы; не поднимет народ на народ меча, и не будут более учиться воевать.

This quotation from the Bible, Isaiah II.4, appears in part as an inscription at the United Nations Plaza. Printed by the Embossograph Process (Thermography) at the 7th Educational Graphic Arts Exposition, New York Coliseum, Sept. 6-12, 1959

FIGS. 1.6 and 1.7   Two examples of previously printed matter.

FIG. 1.8   Previously printed matter composed of text, halftones, and artwork.

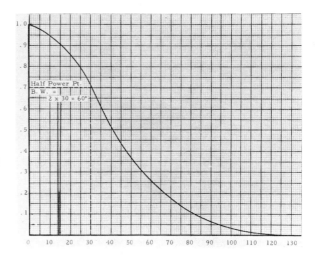

FIG. 1.9   A composite composed of a line original and previously printed matter.

FIG. 1.10   A composite composed of manuscript and previously printed matter.

ink, or brush. Material produced by standard mechanical typewriter is also considered as manuscript.

Previously printed matter (Figs. 1.6 to 1.8) is an original produced by one of the printing processes, such as offset or letterpress. Also included under this classification are black-and-white line photographic prints.

A *composite* (Figs. 1.9 and 1.10) is a combination of two or more of the other forms of line material. Examples include printed business forms containing signatures and engineering records on which pencil notes and ink drawings have been made.

The requirements, methods, and materials for reproducing the five forms of line material by means of a camera are discussed in the remainder of this chapter in the following sections.

Section 1.1 describes the various techniques used in reproducing the five forms of line material; Secs. 1.2 and 1.3 discuss equipment and light-sensitive materials, respectively, needed in camera reproduction of line material; and Sec. 1.4 describes precautions to follow before employing reproduction techniques.

1.1 Reproduction Procedures

An infinite number of problems can arise when reproducing the five types of line material. Some are caused by the type of equipment being used, others by conditions in individual laboratories, and still others are general problems prevailing in all laboratories regardless of their condition or equipment. Discussion of problems caused by individual equipment and laboratory conditions is impossible, since they are local in nature and · must be discovered and overcome by the reproduction technician himself. However, general problems that could prevail in any laboratory are another matter. These are discussed in this section for each form of line material, and solutions are offered that permit satisfactory results at a minimum of time and money expenditures.

Some recommendations may seem to violate theories with which you are familiar. These recommendations are based, however, on experimentation, on years of experience, and on the recommendations of reproduction material manufacturers and leading reproduction firms.

Before discussing specific problems relative to each of the five forms of line material, there are a number of general principles concerning reproduction of all types of line material that should be presented. These are as follows:

General Principle 1. *Use suitable backing on line material.* In copying line material that is on thin or translucent paper, mount it on clean white paper to increase background reflectiveness.

Line material printed on both sides presents a problem if, in copying one side, the printing on the reverse side shows through. To overcome this, mount the copy on a sheet of black paper to provide the same density over the entire background of the side to be reproduced. This situation applies most often when reproducing previously printed matter.

General Principle 2. *Expose and develop for the background, not the image.* This principle violates the long-standing photographic theory of exposing for highlights and developing for shadows. In reproduction, however, we are involved in a photomechanical function and not a photographic process. The procedures, principles, and theories, therefore, often differ. By adherence to this principle, many problems can be eliminated before they arise.

When reproducing line material, the desired result is a negative possessing a *clear image* area against a *dense or black background*. You are not seeking tonal gradations, since original line material possesses no tone—just black and white. Maximum contrast, then, between image and background is essential.

There are times, of course, when every principle must be modified. This one is no exception. Theoretically, black reflects little light and supposedly permits maximum exposure. One assumes, therefore, that maximum background density can be obtained by exposing to the maximum the line material to be reproduced without regard for its black image area.

Actually, the more exposure is increased, the greater will be the reflection bounced back from this black image area—reflection that registers on the film, making the reproduced negative undesirable. You must achieve a mid-point where the negative's background is as dense as possible, but the image area is clear, sharp, and of good quality.

Application of this principle becomes most difficult when reproducing line material having weak pencil or ink lines. You must determine an exposure that provides a negative with a suitable clear image on as dense a background as possible. You may not be able to obtain a background of the desired density, since overexposing the background could lead to obliteration of the image. When working with good original line material, there is no such problem.

Once a camera is properly focused and exposure is made for the background, final image sharpness of the negative depends upon development. Quality reproductions are created when the background goes black within recommended development time, producing a clear, sharp image (Fig. 1.11). Overdevelopment causes loss of detail in the image area (Fig. 1.12), while underdevelopment causes a less-than-sharp image (Fig. 1.13).

The principle of exposing and developing for the background usually should not be violated in reproduction work, except when reproducing reverse originals, that is, an original which has a white image on a black background.

The background of a reverse original is the dominant area, as it is with

THE IMPORTANCE OF

ARCHITECTURAL RENDERING

IN LAYOUT FOR PRESENTATION

FIG. 1.11 Results of a perfectly developed negative. Note that the positive produced from the negative has a sharp image, with all lines uniform, clear, and of the same weight.

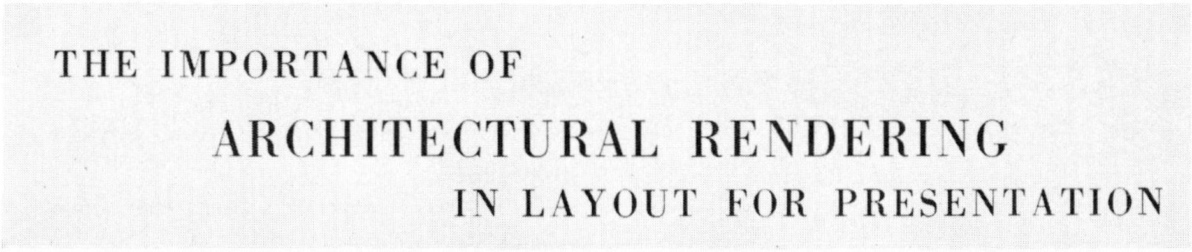

THE IMPORTANCE OF

ARCHITECTURAL RENDERING

IN LAYOUT FOR PRESENTATION

FIG. 1.12 Results of an overdeveloped negative. Note that the positive produced from an overdeveloped negative has a loss of detail in the image, causing loss of fine lines.

FIG. 1.13 Results of an underdeveloped negative. Note that the positive produced from an underdeveloped negative has a less-than-sharp image, causing "spread" around the subject's edges.

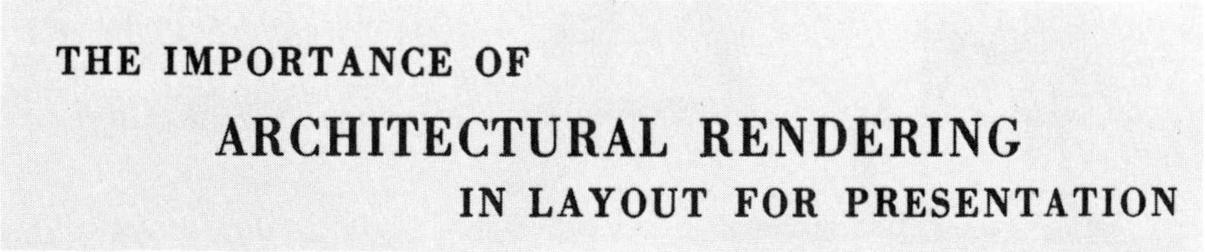

THE IMPORTANCE OF

ARCHITECTURAL RENDERING

IN LAYOUT FOR PRESENTATION

a positive original, because it occupies most of the original's surface. Since a reverse original's background is black, it reflects little or no light. Thus, it requires less exposure than a positive original, which possesses a dominant white background area. In exposing reverse originals, then, it is possible to reduce exposure substantially as compared to that usually required in exposing positive originals.

In developing prints made from negatives of reverse originals, always develop for the white image area. Stop development as soon as this area becomes clear and sharp.

General Principle 3. *Turn off overhead lights.* When making an exposure, all overhead lights must be turned off, especially when exposures are for long durations, as they may be in reproducing line material. Lights could bounce reflections off glass or metal objects in the camera area, causing overexposure of the negative and lens flare.

General Principle 4. *Use orthochromatic film when possible.* (A discussion of light-sensitive materials and their properties follows in Sec. 1.3.) Negatives of line material made with orthochromatic films are easiest to work with during development. You can inspect them frequently under a red safelight, checking for background density and image clearness. This cannot be done with other films.

When inspecting orthochromatic negatives, hold them to the red light (at the recommended safe distance) and look *through* the negative. To determine degree of fine detail, examine the edges of curved letters in their relations to the background (Fig. 1.14). They should be clear and sharp against the background, with no trace of blur, haze, or loss of detail.

General Principle 5. *Use lithographic developers.* Lithographic developers are recommended in reproduction of line material, because they produce maximum image sharpness and excellent background density. When using these, it is extremely important that temperatures be kept within recommended ranges. If too warm, lithographic developers tend to spread the image, making it unclear; if too cold, insufficient background density results.

To summarize, these are the five general principles:

1. *Use suitable backing on line material.*
2. *Expose and develop for the background, not the image.*
3. *Turn off overhead lights.*
4. *Use orthochromatic films when possible.*
5. *Use lithographic developers.*

Keeping these in mind, we now turn to a discussion of how to treat the five types of line material during reproduction.

Line Originals. Line originals are primarily ink or pencil drawings.

FIG. 1.14 Where to check for detail. Arrows indicate those parts of letters that should be carefully checked during development to determine degree of fine detail.

There is little difficulty in reproducing a clear, crisp original; however, technicians are more frequently presented with dirty, worn, and torn drawings. These offer many problems in reproduction.

You should have little trouble reproducing ink originals, in spite of their condition. If ink is chipped or is starting to crack, however, care must be taken when attempting to minimize reflections caused by the marred areas. Stop development immediately when the image area is uniform.

In reproducing bold ink mechanical lettering, carefully inspect the detail of these letters during development. Again, development should be stopped when the image area is uniform and there is maximum contrast between image and background.

Most problems arise when reproducing those line originals drawn in pencil, especially if graphite lines and letters have started to smudge from age and wear. (For a discussion of reproducing intermediates from a pencil original by contact printing, see Chap. 8. Our discussion here is concerned solely with camera techniques.)

If a pencil drawing is solid and dense, treat its reproduction as you would an ink original. If, however, pencil lines are faint because the artist used a hard lead or a light stroke, use diffusers over light sources. Undiffused specular light penetrates the pencil image, producing even weaker lines. (Some types of lighting systems, such as ColorTran, which are discussed later in this chapter, do not require diffusion.)

In the reproduction of faded pencil originals that have little or no contrast with the background, development becomes particularly critical. You must terminate development as soon as the image begins to "block up" (fill in). At this point, an acetic acid stop bath becomes an invaluable tool and must be used immediately to stop development.

The advantages of controlled light sources (see Sec. 1.2) become readily apparent when photographing marred pencil or ink originals. By controlling the light's color quality, you are able to make an adequate reproduction of drawings of the worst quality. This is accomplished by adjusting the light's color to a point where maximum contrast is obtained with the type of film you are using.

Manuscripts. An original manuscript may be prepared with pencil, ink, or standard typewriter. Low image-density problems, similar in nature to those encountered with line pencil originals, can occur in reproduction.

In reproducing a manuscript, it is necessary first to evaluate its most critical area. During development of the negative, inspect this area carefully to ascertain the instant maximum image clearness and background density have been reached.

Every method at your disposal for increasing contrast between background and image must be used when reproducing manuscripts or, for that matter, any other type of line material. If the paper on which an original manuscript appears, for example, is off-white or yellowish, use of a yellow filter will improve separation. If the image is very weak, reducing exposure and increasing development will enhance contrast.

When using a high-contrast developer to reproduce a manuscript (or line original) possessing a weak image, it is often necessary to let the image develop out slowly. Be certain, however, that you watch closely to determine the moment image and background have reached maximum contrast proportions.

Weak images are best reproduced on paper-base negative materials rather than on acetate-base. One disadvantage of the former, however, is that they are impractical for platemaking. Paper-base materials require a lengthy exposure in the printing frame preparatory to making a plate. Although it can be done, the time required for light to pass through the paper material may not be justified.

Paper-base materials, however, are extremely useful in making intermediates for such reproduction methods as diazo and blueprint (see Chap. 8). When developing this type of negative, it is important to remember one thing: inspection of the negative during development is accomplished by looking through its *back*. Only in this way can you obtain an accurate picture of the image-background contrast.

When reproducing typewritten or ink manuscripts with orthochromatic film, take note of the color of the image characters. If they are blue or purple, you may encounter difficulty in making the reproduction, since orthochromatic film tends to "lose" blue.

To overcome this, change the color of blue characters before reproduction by fuming the manuscript over an open bottle of ammonia, which converts characters to brownish-black. Purple ink can be changed to black by fuming a manuscript with sodium sulfide or by covering the area with a blotter saturated with a concentrate of this solution. You may also darken blue and purple ink characters by use of a deep yellow filter.

Previously Printed Matter. An inexhaustible source of reference and research information for engineering personnel are technical articles, books, and pamphlets with their many halftones, sketches, and diagrams. These are usually printed by the offset or letterpress process. Some technicians claim that reproduction of this source material for distribution is their primary responsibility.

Before reproducing previously printed matter, make sure you are not violating copyright laws. If necessary, secure written permission of the author or publisher authorizing you to reproduce part or all of an entire document.

When previously printed and illustrated matter is being reproduced, the objective is to maintain the sharpness and clarity of the original. This is often difficult to do, especially when reproducing halftones, but it is possible by employing one of a number of methods, such as diazo (whiteprint), blueprint, photographic printing, or offset printing.

Whichever method you use, the reproduction can be as good as the original previously printed matter if extreme caution is taken during the first two steps of the process. No matter which method of reproduction you will eventually use to make final copies, the procedures for these two steps are much the same. The final step of actually making the copies depends, of course, on the particular technique you wish to employ or the equipment available.

Step 1. Immediately, your main concern is preservation of the original's intricate detail and clarity. It is best to copy previously printed matter with a graphic-arts reproduction camera. In its absence, use the largest view camera available in your laboratory.

If possible, shoot a negative the same size as the original copy to maintain the separation required in fine detail areas, particularly of halftones. Shooting a negative the same size as the original is usually accomplished easily with any type of reproduction camera. If only a view camera is available, however, enlarging after you obtain a negative may be necessary for reproduction by diazo, blueprint, or photographic printing. If the negative is smaller than the required reproduction, the negative will have to be placed in an enlarger and a film or paper positive of the necessary size made. Only large-size view cameras (8 by 10 in. and larger) produce adequate image sizes for production of negatives for offset printing plates.

Extreme care in focusing is essential, particularly if the final reproduction negative master is made with a view camera and must be enlarged. Focusing should be checked in the most critical area. When reproducing halftones, this area is the dot structure. (See Chap. 2 for a discussion concerning the reproduction of halftones.)

Make exposures on a contrast-process lithographic film for maximum contrast and background density. These films permit inspection under a red safelight during development. If reproducing text and halftones on the same negative, be extremely careful to reproduce both with equal clarity.

Step 2. Development of the exposed negative is as critical to reproduction quality as exposure and lighting. Use the fine-line development technique, which is explained below. This method brings out sharply the fine-dot structure of a halftone and provides the excellent-image character of the original copy. (Fine-line development is also used when reproducing other types of line material. In certain situations discussed throughout this book, it is indispensable in achieving sharp reproductions.)

Fine-line development consists of developing a negative in a high-contrast lithographic developer, such as Eastman Kodak's fine-line development powder, using brief agitation until the image becomes visible. *Still development* is then employed for the remainder of the recommended development time (Fig. 1.15). Total average fine-line development time is about 2¼ min. This time is critical, since overdevelopment veils a negative's clear areas.

FIG. 1.15 The fine-line development technique. Still development (no agitation) is used after the image becomes visible.

Line Duplicates. Line duplicates (such as blueprints, photostats, sepia drawings, blueline diazo prints, and types produced by the many kinds of office copying machines) can be extremely difficult to reproduce. During those infrequent times when line originals are not available or are not suitable, and you must make acceptable reproductions from duplicates of these originals, remember the important principles of *filtration, exposure,* and *development.*

Occasions may arise (copying an old, faded photostat is one) where the duplicate is extremely weak. As a last resort, use a long-scale continuous-tone film to try to record the image. Develop the negative obtained from this commercial-type blue-sensitive film in a high-energy developer. Even developers such as DK:60 and Isodol are suitable.

Following development, print the negative on the hardest-grade photographic paper available in your laboratory. The developed print is usually superior in quality to the line duplicate you started with. If a reproduction negative is then needed, make it by copying this print.

When you are requested to reproduce a faded, mangled illegible blueprint, it is important to determine immediately to what use the reproduction will be put. Will it be used for distribution to shop or technical personnel? Or will it be filed for record? Your answer determines the reproduction method to employ.

If for file purposes only, maintaining fine detail may not be important. Thus, you can copy the duplicate photographically with commercial-type film and make a print. If, on the other hand, the reproduced blueprint is going to be used by shop personnel and technicians in their work, it is imperative to maintain as much detail as possible. To accomplish this, you must make a negative intermediate from the poor original, from which copies can be made by either the blueprint or diazo method.

To make an intermediate from an original having sufficient contrast, reproduce the original on either a short-scale contrast-process panchromatic or orthochromatic film. Use a red filter with panchromatic film and a yellow filter with orthochromatic. Each filter used with its respective film darkens the blue background and gives a clearer white image, thereby enhancing contrast (Fig. 1.16).

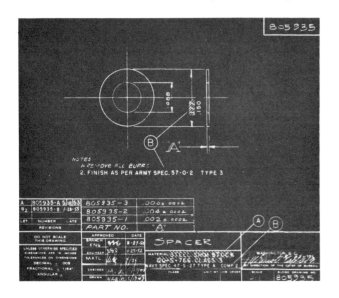

FIG. 1.16 Reproducing a contrasty blueprint. This reproduction was made from the original blueprint with short-scale contrast-process panchromatic film and a yellow filter.

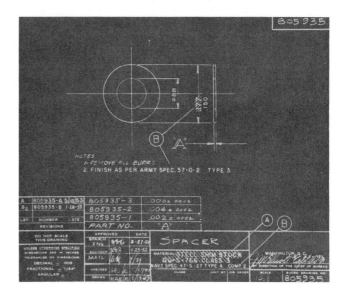

FIG. 1.17 Reproducing a faded blueprint. This reproduction was made with a long-scale continuous-tone high-contrast panchromatic film and a red filter to darken the background.

If panchromatic film is used, develop in total darkness. A dark green safelight can be used from time to time for inspection, but only as recommended on the film package. It would be practical to standardize your procedure when using this film by performing a series of exposure tests on several blueprint originals having varied degrees of contrast. After these tests have been made, you can adjust exposure according to the type of original being reproduced. When processing in the dark, the procedure would then be the same in all cases since contrast of the developed negative is determined by exposure and not by development. The green safelight can be used only occasionally to check degree of development.

Remember one point if you are tempted to employ the green safelight more than is recommended: green is used as a safelight color because your eye is most sensitive to this color, and not because the film is least sensitive to it.

As pointed out before, there is no difficulty in inspecting orthochromatic film in the darkroom. You may check this film continuously under a red safelight, provided the film is held at the recommended distance from the safelight.

If the original blueprint for reproduction is faded, you may not be able to improve on it a great deal by using contrast-process films and their respective filters. In such cases, reproduce the image on a long-scale continuous-tone film. Develop in a high-contrast developer, increasing development time to obtain as much contrast as possible.

This method is used as a last resort. The continuous-tone negative, although providing less contrast than desired, may produce a legible image that can be printed by standard photographic techniques (Fig. 1.17).

Composites. In reproducing composites, where more than one of the five types of line material are involved, first determine whether the reproduction is for record purposes only or whether the reproduction is for distribution. If for distribution, what reproduction method will be employed; that is, are many copies needed, necessitating a printing process, or are only a limited number of copies needed, making the use of a photographic process suitable? Finally, determine whether the entire composite is of importance or whether only a limited area has to be reproduced. This determination may save a great deal of work if a composite is large and of poor quality. It is a waste of time and energy to attempt reproduction of an entire composite when only a small portion of it is desired.

In reproducing composite originals, weak areas are of prime importance. If detail in weak areas can be preserved, detail in solid areas will also be preserved.

It may not be possible to reproduce a composite in which there are extreme differences in image densities on short-scale process film. The solution is to photograph the original composite on a long-scale continuous-tone (blue-sensitive) commercial film and develop for maximum contrast. The result is a high-contrast continuous-tone negative, which makes possible photographic printing. This method may be the only way of reproducing a poor-quality composite so all detail is maintained.

Four basic rules to observe when reproducing composite originals are as follows:

1. If the composite has important areas (signatures, for example) that are poor in quality, expose and develop for these areas.

2. In reproducing composites that have color areas, use a filter that brings out main areas or one that provides maximum color separation. Be certain the filter you select does not completely filter out a required area.

3. If reproduction of fine detail is important, use the fine-line development technique discussed earlier.

4. Check development carefully. Stop it as soon as the most critical area begins to "block-up."

1.2 Equipment

Quality reproduction of line material depends on the way you employ your equipment and not necessarily on the type of equipment you employ. The expert technician is one who is thoroughly familiar with the characteristics and limitations of each piece of equipment with which he works.

This section discusses the most important equipment found in a reproduction laboratory: camera, lens, copyboard, lighting, and filters. It is intended to acquaint you with the general characteristics and uses of each. Once familiar with these, however, the technician should examine and test

FIG. 1.18 Graphic-arts reproduction camera and copyboard. (*Courtesy of Lockheed Electronics Company.*)

FIG. 1.19 Graphic-arts reproduction camera, overhead model. (*Courtesy of R. W. Borrowdale Company.*)

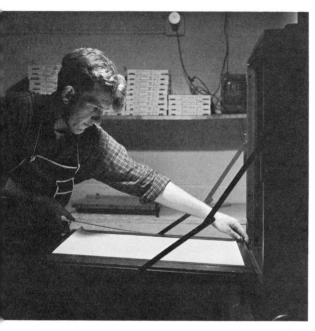

FIG. 1.20 Loading film into the back of a 20 by 24 in. graphic-arts reproduction camera. (*Courtesy of Lockheed Electronics Company.*)

his particular equipment, since each has its own characteristics. For that matter, two pieces of the same equipment made by one manufacturer have their own distinct peculiarities.

The Camera. Practically any camera can be used in reproducing line material, provided the basic principles of good reproduction are followed. Our discussion, however, is confined to professional cameras used by reproduction engineers, commercial photographers, and others engaged in reproduction work. These are graphic-arts reproduction cameras, view cameras, and press-type cameras.

Graphic-arts reproduction cameras (Fig. 1.18) are specially designed and engineered to provide accurate alignment between copyboard and film plane. They are large pieces of equipment, available in a variety of types and sizes (Fig. 1.19).

Graphic-arts reproduction cameras hold film the size of which corresponds to the size of the camera; that is, the usual 20- by 24-in. reproduction camera holds sheet film up to but not exceeding this size (Fig. 1.20).

Facilities employing graphic-arts reproduction cameras are those usually

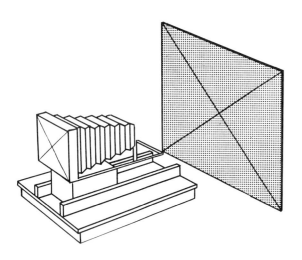

FIG. 1.21   A typical view camera setup.

engaged in volume reproduction, such as lithographic, engraving, and engineering reproduction companies.

View cameras (Fig. 1.21) are also employed by many facilities. There are two general types: (1) those designed specifically for copying; (2) the standard view camera, which is used primarily as an all-purpose photographic camera, but which can also be employed for copying.

View cameras most desirable for reproduction work are those that have moving front and rear standards; that is, a movable lens and film plane. The front standard is used for critical focusing, while the rear standard is used to obtain correct image size.

View cameras are available in sizes of 11 by 14 in., 8 by 10 in., 5 by 7 in., and 4 by 5 in. They produce excellent reproductions of line material, provided they are equipped with strong stands or supports to hold them steady during operation.

Another camera that can be used on occasion for reproduction is the *all-purpose camera:* the 4- by 5-in. press camera and other smaller-size handheld equipment. These are known for their performance in conventional photographic work, but can be employed by photography firms that receive an occasional order to reproduce a piece of line material. The expense involved in purchasing a reproduction or view camera for this infrequently received work may not, of course, be justified.

The Lens. If we desire a general rule about lenses, it would be this: lens quality determines reproduction quality. *But the most expensive lens does not compensate for mechanical or human error in the copying setup.* If the copyboard and film plane are not parallel, for example, lenses having the greatest precision would not correct the distortion that results.

You can use practically any lenses for reproducing line material, even those without a shutter, which are usually found in older-model view cameras. When a lens is used in reproducing line material, it is usually open for an extended time to permit proper exposure. If a shutterless lens is used, this lengthy period requires employment of a lens cap. Simply remove the cap from the lens at the beginning of exposure, and replace it at the end of exposure.

Most graphic-arts reproduction camera lenses are equipped with electrically operated shutters. The *process anastigmat lens* is the one recognized by reproduction technicians as producing reproductions of the greatest sharpness.

Before a new lens is used for the first time, it should be tested to determine the aperture opening (f/stop) at which the sharpest reproduction is produced. To arrive at this critical point of resolution, set up a piece of previously printed matter, such as a newspaper or magazine page, for copying and make an exposure a each f/stop. You may be surprised to find that the smallest aperture opening does not always produce the sharpest image, as it does in conventional photography. In reproduction, most lenses reach their critical point of resolution at two or three stops below the *widest* opening. Remember, however, that in reproduction work focus is on a flat plane. Consequently, depth of field is never a consideration, and it matters little how wide or small the lens aperture opening. Always select that opening which provides the sharpest reproduction.

The reproduction technician is concerned with reproducing an image either the same size as a piece of line material, or smaller (reduction), or larger (enlargement). Most graphic-arts reproduction cameras are designed with easy-to-operate slide-rule type calculators which permit determination of the exact placement of lens and copyboard for same size reproduction, reduction, and enlargement.

Suppose, however, that a graphic-arts camera with its built-in scales is not available and you must work with another type of camera which has a fixed copyboard. How do you determine correct lens-to-subject and lens-to-film distances without employing time-consuming, complicated mathematical calculations? The formulas below answer this question. If you employ them properly, you will have no trouble finding correct distances and will have no need for other mathematical manipulations.

Before using these formulas, familiarize yourself with the following key:

F = focal length of lens
h = height of subject matter
h^1 = height of focal-plane image
m = magnification
u = lens-to-subject distance
v = lens-to-film distance

In reproduction, known factors are focal length of the lens (F), height of subject matter (h), and height of image desired (h^1). You may or may not know magnification (m) of the matter to be reproduced. The best way to determine magnification, if it is not readily apparent, is with a photographer's proportion rule (Fig. 1.22). This piece of inexpensive equipment is invaluable when doing reproduction and photographic work.

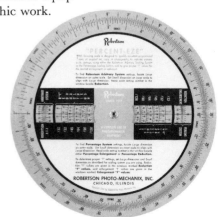

FIG. 1.22 Calculator for determining percentage of reduction, enlargement, and image proportions. (*Courtesy of Robertson Photo-Mechanix, Inc.*)

Factors not known are lens-to-subject distance (u) and lens-to-film distance (v), but these can be determined by the following formulas:

Formula 1—To Determine Lens-to-film Distance v:

$$v = (m + 1) \times F$$

EXAMPLE. Determine lens-to-film distance (v) when reproducing a piece of line material the same size [a magnification (m) of 1:1]. Focal length of the lens (F) is 7½ in.

$$
\begin{aligned}
v &= (m + 1) \times F \\
&= (1 + 1) \times 7\tfrac{1}{2} \text{ in.} \\
&= 2 \times 7\tfrac{1}{2} \text{ in.} \\
&= 15 \text{ in.}
\end{aligned}
$$

Set the distance between lens and focal-plane image (film-plane) by adjusting the bellows extension 15 in.

Formula 2—To Determine Lens-to-subject Distance u:

$$u = \left(\frac{1}{m} + 1\right) \times F$$

EXAMPLE. Determine lens-to-subject distance (u) when reproducing a piece of line material the same size [a magnification (m) of 1:1]. The lens has a focal length (F) of 7½ in.

$$
\begin{aligned}
u &= \left(\frac{1}{m} + 1\right) \times F \\
&= \left(\frac{1}{1} + 1\right) \times 7\tfrac{1}{2} \text{ in.} \\
&= (1 + 1) \times 7\tfrac{1}{2} \text{ in.} \\
&= 2 \times 7\tfrac{1}{2} \text{ in.} \\
&= 15 \text{ in.}
\end{aligned}
$$

Set the distance between lens and subject at 15 in.

Lens problems of the type depicted by these formulas always require employment of both calculations at the same time. These problems are often more complicated than those cited above. The following, for example, is a typical one.

Reproduce a 40- by 60-in. subject area to one-tenth its size. Focal length of the lens is 7½ in.

To Determine Lens-to-subject Distance u:

$$
\begin{aligned}
u &= \left(\frac{1}{m} + 1\right) \times F \\
&= \left(\frac{1}{\frac{1}{10}} + 1\right) \times 7\tfrac{1}{2} \text{ in.} \\
&= (10 + 1) \times 7\tfrac{1}{2} \text{ in.} \\
&= 82\tfrac{1}{2} \text{ in.}
\end{aligned}
$$

To Determine Lens-to-film Distance v:

$$v = (m + 1) \times F$$
$$= (\tfrac{1}{10} + 1) \times 7\tfrac{1}{2} \text{ in.}$$
$$= (1\tfrac{1}{10}) \times 7\tfrac{1}{2} \text{ in.}$$
$$= 8\tfrac{1}{4} \text{ in.}$$

Set the subject 82½ in. from the lens, and extend the bellows 8¼ in. from the focal-plane image (film-plane) to reproduce a 40- by 60-in. subject area to one-tenth its size.

Another lens problem inherent to reproduction work is loss of light occurring because of long bellows extensions. The longer the bellows extension, the weaker light becomes as it travels from lens to film plane. Allowances for this loss must be made or reproduction will be less than perfect.

A rule to remember is: *the strength of light varies inversely as the square of the distance.* For every inch light travels in the bellows, it becomes twice as weak. Moreover, when a camera's lens is *closer* to the subject than eight times the focal length of the lens, compensation must be made for loss of light inside the bellows or underexposure results. If lens-to-subject distance is greater than eight times the focal length of the lens, normal exposure meter readings can be safely used.

To compensate for light loss inside a bellows, the formula to employ is

$$\frac{\text{BE}^2 \text{ (bellows extension)}}{\text{FL}^2 \text{ (focal length)}} = \text{factor to multiply by meter reading}$$

Determining Exposure. The data concerning lens distance and compensation for light loss inside a bellows are of little value unless the basic procedure for determining proper exposure is known. Proper exposure is best determined by making a series of test negatives immediately after establishing a copy setup. Thereafter, all exposures can be standardized. This method is applicable to both line and continuous-tone material.

To standardize exposures by means of the test negative method, first determine the exposure at which to begin. Use a photoelectric exposure meter to obtain an approximate exposure.

Never take a meter reading directly from the copy in the copy easel. Instead use a neutral test card, which reflects 18 per cent of the light received. This gray-colored card, referred to as Kodak 18 per cent reflectiveness gray card, is manufactured by Eastman Kodak Company. If one is not available, however, you can use a sheet of white paper from which to obtain a meter reading and then *divide this reading by 5.*

After obtaining the meter reading, make the test exposure or a series of exposures at varying shutter speeds and at the most suitable lens opening (as we said before, the most suitable opening is usually two or three stops below the widest opening). By developing and examining these negatives, you can determine the one which provides the desired density or the adjustments that should be made to obtain this desired density.

The Copyboard. Graphic-arts reproduction cameras are usually equipped with built-in easels that hold the original while copying (Fig. 1.23). These are often glass-enclosed tilting frames that keep the entire surface of a

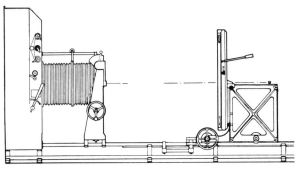

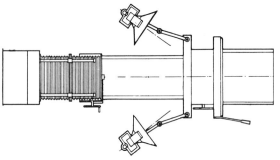

FIG. 1.23 Relationship of lens to copyboard in a graphic-arts reproduction camera setup. Both front and rear standards are movable to obtain the desired scale. (*Courtesy of Chemco Camera Company.*)

piece of line material under equal pressure. Many reproduction facilities also employ vacuum easels that are equipped with vacuum pumps. These hold the original flat against the easel by suction.

A makeshift copyboard can be used if other than a graphic-arts reproduction camera is employed. Any arrangement, even a flat piece of board, is suitable, provided it meets the following conditions:

1. The copyboard must be parallel to the film plane and level to prevent distortion. Employ a carpenter's level to ensure both conditions. First check the level of the copyboard itself; then check the film plane for parallel alignment.

2. If the copyboard is made of wood, it should be painted flat black to prevent glare. Never use gloss paint. A copyboard made of cork is excellent and simplifies the task of tacking the original to the board. It, too, should be painted black.

3. The copyboard must be large enough to accommodate the largest original you will reproduce in one section. Make sure the board's entire area is taken in by the covering field of the camera's lens. There are cases in which the original is reproduced in sections and pieced (stripped) together. Sectional negatives and how to handle them in stripping are discussed in Chap. 5.

Lighting. Light is a vital tool in reproduction and must be carefully considered. The type and quality of illumination always affect the quality of the finished product.

Light sources for reproduction, discussed here, have been divided into two categories for the sake of clarity. This has been done to place emphasis upon those sources usually employed with graphic-arts reproduction cameras. Other sources relative to photography are mentioned in passing and some general, but important, principles are given for their application.

Keep in mind that *light sources can be interchanged,* no matter what type of camera is used.

The two classifications of illumination sources are: (1) general sources, such as fluorescent lighting, standard incandescent illumination, and photo-engravers' lamps, which are usually employed with cameras other than re-production cameras; (2) specialized sources, such as carbon-arc lamps and those permitting control of the light's quality (ColorTran converters and pulsed xenon-arc lights, for example), which are used primarily with graphic-arts reproduction cameras.

When *general light sources* are used with view or small-size cameras, the copyboard should be illuminated as evenly as possible. Acceptable lighting is obtained by placing lights in a fixed position, 45° from the lens axis on each side of the copyboard. Measure the distance between each light and the center of the copyboard to ensure evenly distributed illumination. *Each light must be exactly 45° from the axis of the lens on each side of the copyboard; each must be the same distance from the copyboard.*

In the absence of these requirements, even lighting will not be obtained, and line material possessing faint or light image lines will result in neg-atives that are difficult to print. Some areas will be overexposed; others will be underexposed. To further ensure even lighting, use reflectors of the same design and bulbs of the same rating.

If you are reproducing black-and-white material only, it is not necessary to replace all bulbs in an incandescent-lighting system when one needs to be changed, with the exception of the ColorTran system. With ColorTran, you should change all bulbs when one burns out to maintain the same Kelvin temperature throughout, thus ensuring complete control over the spectral range. If working with color film, you should replace all bulbs in *any* type of lighting system with new ones when one burns out. Thus, cor-rect color balance is maintained. (For a discussion of Kelvin temperature and color balance, see Sec. 5.5.)

A simple check can be made to ensure uniformity of light with general illumination sources. Put a sheet of white paper in a copyboard. Place the edge of a pencil or ruler at a 90° angle (perpendicular) over one area of the paper and then over another. Compare the shadows cast on each side of the pencil or ruler for uniformity. If one side is darker than the other, a light is out of position. Adjust that light until shadows are of equal density. Although not exceptionally accurate, this test does permit the making of simple adjustments.

The following *specialized light sources* are usually employed with graphic-arts reproduction cameras.

1. *Carbon-arc lamps,* a standard method of illumination that has been in use for many years, provide high-intensity illumination at low exposure times. Caution must be exercised to guard against overexposing line material of poor quality, since the penetrating power of these lamps is great.

Carbon arcs operate at a constant voltage which, in most cases, cannot be lowered to provide a less-intense light or a reduction in color tempera-ture. Compensation for the highly intense carbon-arc light is obtained, however, by decreasing exposure and controlling development.

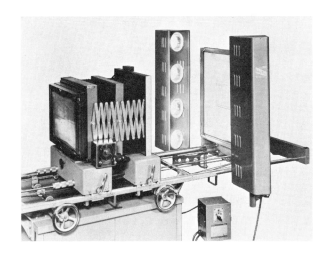

FIG. 1.24 ColorTran 30° lighting setup. (*Courtesy of Natural Lighting Corporation.*)

Carbon arcs are efficient light sources when connected to electric lines that provide an almost constant voltage, with no great deviation. When reproducing line material of good quality, voltage fluctuation may not present a problem since it may be compensated for during development. Difficulties arise, however, when voltage fluctuates as critical exposures are being made, such as when reproducing color line material on color film. A relatively minor fluctuation in voltage of just 1 volt can result in a 10° drop in Kelvin temperature. To guard against difficulties such as this, have line voltage checked periodically to determine amount of fluctuation. When doing critical work, you will then have some idea of limitations and pitfalls.

2. *ColorTran* light (Fig. 1.24) eliminates most of the shortcomings encountered with carbon-arc lamps. An outstanding feature of ColorTran

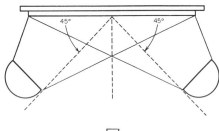

FIG. 1.25 Typical camera lighting setup with lights at a 45° angle to the copyboard.

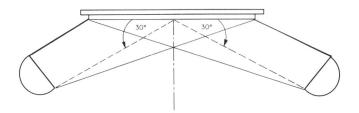

FIG. 1.26 ColorTran lighting setup with lights at a 30° angle to the copyboard.

is the evenness of illumination it provides at the film plane, regardless of the copyboard's size.

"Even" illumination (also called *flat* lighting) with other lighting systems, including carbon arcs, is not an accurate description. With these, the copyboard's central area is illuminated 25 to 35 per cent more than the outer edges. As light reaches the film plane, the center section of the material for reproduction receives more exposure than the outer edges. Compensation is made by developing until the outer edges reach the density of the central portion. When reproducing line material of poor quality, this procedure often leads to overexposure or a partially veiled image in the central area.

With ColorTran, however, lighting units are positioned at a 30 to 35° angle rather than at a 45° angle demanded by other lighting systems, including carbon arcs (Fig. 1.25). Light reaching the film plane is compensated and of equal density at the edges as in the center, thus alleviating manipulation in development (Fig. 1.26).

ColorTran offers another feature not characteristic of other lighting systems: light intensity can be controlled by adjusting the spectral range, permitting a technician to select the correct light quality for the copy being rendered. With high-intensity carbon arcs, light pencil drawings are easily lost. By using a ColorTran converter, a technician can reduce light intensity and build up contrast by adding more yellow light, making it possible to hold lines that are thin and weak. This feature provides an operator with a latitude of from one to three gray stops more than with carbon arcs.

3. *Pulsed xenon lights* (Fig. 1.27) are relatively new illumination sources. Features include constant daylight color over most of the spectral range, constant light output, and relatively cool and clean burning lights that help to minimize dust.

Pulsed xenon lights are lamps filled with low-pressure xenon gas. Their power is almost 100 watts per in., with the total lamp length determining power; that is, total power of a lamp is arrived at by multiplying 100 watts per in. by the overall length of the lamp.

These lights provide full daylight color over most of the visible spectrum, from 400 to 700 millimicrons, offering a medium response as far down in the ultraviolet range as 300 millimicrons. Light quality, in effect, can be

FIG. 1.27   Pulsed xenon lighting setup. (*Courtesy of American Speedlight Corporation.*)

ORIGINAL COPY	to photograph as	when using Orthochromatic FILTER	when using Panchromatic FILTER	when using Blue Sensitive FILTER
BLUE	Black	**YELLOW** or **ORANGE**	**ORANGE** or **RED**	use **ORTHO** film
	White	**BLUE**	**BLUE**	**NONE** or **BLUE**
BLUE–GREEN	●	use **PAN** film	**RED**	use **PAN** film
	○	**BLUE** or **GREEN**	**BLUE** or **GREEN**	use **ORTHO** film
GREEN	●	use **PAN** film	**RED**	use **PAN** film
	○	**GREEN** or **YELLOW**	**GREEN**	use **ORTHO** film
YELLOW–GREEN	●	**BLUE**	**BLUE**	**NONE** or **BLUE**
	○	**GREEN** or **YELLOW**	**GREEN**	use **ORTHO** film
YELLOW	●	**BLUE**	**BLUE**	**NONE** or **BLUE**
	○	**YELLOW**	**RED, ORANGE** or **YELLOW**	use **ORTHO** film
ORANGE	●	**BLUE** or **BLUE-GREEN**	**BLUE** or **BLUE-GREEN**	**NONE**
	○	use **PAN** film	**RED** or **ORANGE**	use **PAN** film
RED–ORANGE	●	**BLUE** or **BLUE-GREEN**	**BLUE** or **BLUE-GREEN**	**NONE**
	○	use **PAN** film	**RED** or **ORANGE**	use **PAN** film
RED	●	**NONE, GREEN** or **BLUE**	**GREEN**	**NONE**
	○	use **PAN** film	**RED** or **ORANGE**	use **PAN** film
VIOLET	●	**DEEP YELLOW**	**GREEN**	use **ORTHO** film
	○	**BLUE**	**BLUE** or **MAGENTA**	**NONE** or **BLUE**

FIG. 1.28 Filter chart for graphic-arts films. (Courtesy of E. I. du Pont de Nemours and Company, Inc.)

considered close to that of sunlight, giving excellent results when used with color and black-and-white line material. They can be started and stopped instantly at full intensity and at a constant color balance.

Filters. Figure 1.28 provides in chart form valuable information concerning the use of filters in line reproduction. In reproduction of black-and-white line material, filters can be employed to reduce the effects of or to eliminate entirely an objectionable background. For example, when reproducing a diazo print that was underexposed, making the background appear blue, use of a blue filter would lighten this tinge. Filters are also employed in the reproduction of color line material to darken or lighten colors for better contrast.

A basic rule of thumb to follow when working with filters is: *any one of the three basic filters—red, green, and blue—lightens its own color and darkens the other colors.* Blue filters out blue and darkens red and green; red filters out red and darkens blue and green; and green filters out green and darkens blue and red.

Additional Accessories. In addition to equipment already discussed in this section, there are other accessories that aid in the reproduction of line material. This equipment is as follows:

1. A 10-power magnifier for checking fine detail and making accurate measurements. Magnifiers for reproduction are available from graphic-arts, photographic, and optical suppliers.

2. A timer or stopwatch for timing exposures made on cameras having nonautomatic shutters.

3. A carpenter's level for checking alignment of those copy setups usually required when other than a graphic-arts reproduction camera is employed.

4. A transparent ruler for making measurements on a camera's ground glass.

5. Black-and-white paper for backing up line material.

6. A metal straightedge for cutting and stripping line negatives.

7. Opaquing solution (black or red waterproof opaque is easiest to handle with line negatives).

8. Opaquing brushes (medium-bristle brushes are preferred for spotting).

9. Masking tapes for blocking out and stripping sections together (masking tapes are available from graphic-arts suppliers in a variety of colors that hold back light in exposure, but permit you to see through them while stripping).

10. Rubber cement, clear tape, thumbtacks, safety-razor blades, hard-lead pencils (No. 4 or 5), and gum erasers. These items are available from general stationery suppliers or other consumer product sources.

11. A light table for viewing, spotting, and stripping (this essential tool can easily be fabricated at moderate cost). Many types of light tables are available. These range from simple units to complex lineup tables with microsettings.

1.3 Light-sensitive Materials

Familiarity with light-sensitive materials and selection of the one best suited for a particular task are as important to the creation of quality reproductions as any other factor (Fig. 1.29). Fortunately, light-sensitive materials available for reproduction of line material in the camera are limited to a few basic types. Do not be confused by the many products sold. They are just variations of these types, differing only in emulsion speed, the support or base on which the emulsion is suspended, and color sensitivity.

Before discussing films specifically manufactured for reproduction of line material in the camera, let us examine briefly the general area of films. Films are divided into short-scale and long-scale materials. Short-scale films have little latitude. They "see" only black and white and do *not* produce continuous tones. Conversely, long-scale films have considerable latitude, making them more flexible in exposure. They are continuous-tone materials capable of producing gradations of the gray scale and are used mainly for conventional photographic work.

Short-scale films are the ones used in reproducing line material. They provide excellent contrast, a prime requisite for high-quality reproduction, and are divided into two general categories: orthochromatic and panchromatic. When you hear or read the terms *photomechanical, process,* and *contrast process,* realize that they are being used in describing short-scale films.

Orthochromatic films, which are particularly sensitive to blue and green, are divided into two types, as follows:

1. *Contrast-process orthochromatic film,* a professional material that offers excellent contrast and fine grain, is sensitive to blue and green. It is approximately eight times faster than lithographic orthochromatic film. Contrast-process orthochromatic material is available only on a regular acetate base and is generally not suited for ultrafine-line reproduction by contact printing and offset platemaking.

2. *Lithographic (or graphic-arts) orthochromatic film* is less sensitive to blue and green than contrast-process orthochromatic, but produces much sharper images because of its increased resolution and thinner base. Specifically designed for lithographic use, it is known for its ease of processing with lithographic developers.

Since lithographic orthochromatic film is available in many sizes and in many types of bases, it is considered the best film to employ in reproducing line material. Added features include ease of handling in stripping, competitive market prices, and brief drying times (as short as 4 min with a heater). Lithographic orthochromatic films are available in glass plates, paper base, acetate base, and dimensionally stable base.

Short-scale panchromatic films, which are fairly sensitive to all colors, are also divided into two types, as follows:

1. *Contrast-process panchromatic film,* a professional sheet film available in standard-cut sizes and on acetate base, produces high-contrast, fine-grain reproductions of line material. Like contrast-process orthochromatic film, process panchromatic does not produce ultrafine-line qualities required for critical reproduction work. Its speed is about four times faster than lithographic panchromatic film.

2. *Lithographic (or graphic-arts) panchromatic film* produces very high contrast reproductions and is more flexible to handle than process panchromatic because of its thin film base. Many of the physical characteristics applicable

SILVER SENSITIVE MATERIALS

FOR THE REPRO FIELD

Producers ▶
◀ Products

	AMERICAN PHOTOCOPY	ANKEN	ANSCO	BRUNING	CHEMCO	G. CRAMER	DIETZGEN	DI NOC	DU PONT	KODAK	GEVAERT	GRANT	HUNTER	KEUFFEL & ESSER	KILBORN	PEERLESS
PAPERS & VELLUMS																
Contact, Room Light Negative or Print	X						X			X		X				X
Contact, Room Light Positive-to-Positive		X					X			X	X	X	X			X
Contact, Darkroom	X	X	X	X			X			X	X	X	X	X		X
Projection-Medium		X	X	X			X		X	X	X	X	X	X	X	X
Projection-Fast		X	X				X		X	X	X	X	X			X
Projection-Litho									X	X	X	X	X			
FILMS (ACETATE, TRIACETATE, VINYL)																
Room Light Positive-to-Positive		X								X	X	X				
Contact, Darkroom Negative-Clear	X	X		X	X		X					X	X	X		
Contact, Darkroom White Base		X					X					X		X		
Projection, Litho			X	X	X		X	X	X	X	X					
FILMS (RELATIVELY STABLE)																
Room Light Positive-to-Positive										X	X	X	X			
Contact Darkroom Positive-to-Positive		X								X		X	X	X		
Projection Positive-to-Positive			X									X				
Projection, Litho White:									X	X	X		X			
CLOTH — GLASS																
Contact Darkroom Negative							X					X		X		
Contact Darkroom Reflex Negative							X					X		X		
CLOTH — LINEN																
Contact Darkroom White				X			X					X	X	X		X
Contact Darkroom Blue				X			X					X	X	X		X
Room Light Wash-Off				X			X					X	X	X		
Room Light Negative												X	X			
Room Light Positive-to-Positive												X				
Projection				X			X					X	X	X		X
MISCELLANEOUS																
Adhesive Papers Pressure Sensitive			X		X							X		X		
Template Emulsion		X	X			X		X	X			X		X		

FIG. 1.29   Silver-sensitive materials for the reproduction industry. (*Reprinted by permission of Industrial Photography Magazine.*)

to lithographic orthochromatic film also apply to lithographic panchromatic film, which is available in glass plates, acetate base, and dimensionally stable base.

Most short-scale orthochromatic and panchromatic lithographic films are available only in sizes of 8 by 10 in. and larger. Most short-scale orthochromatic and panchromatic process films are available in these sizes as well, but also in sizes as small as 4 by 5 in.

When considering which film to use, examine your requirements in light of the following:

Contrast Desired. For maximum contrast, lithographic orthochromatic or panchromatic films are best; for moderate contrast, use contrast-process orthochromatic or panchromatic film.

Color Sensitivity. Orthochromatic films are particularly sensitive to blue and green. Panchromatic films are fairly sensitive to all colors, but can be viewed under a green safelight for short periods when developing. Green is used because the eye is most sensitive to it, permitting inspection to be accomplished in a minimum of time.

1.4 Precautions in Preparation for Reproduction

The painstaking and meticulous day-to-day work performed by the technician to meet the critical demands made upon him in reproduction is hampered by a natural phenomenon—dust—which he can control, but cannot eliminate. These ever-present particles can cause incalculable losses in time, money, and effort unless the reproduction technician does everything possible to minimize their effects.

Modern advances in reproduction technology have brought home forcefully the damage dust can do in a laboratory. Introduction of new high-resolution lenses, films and papers, and printing mediums essential for the creation of the highest-quality work only hamper the technician in his constant struggle against dust. When requested to reproduce a piece of line material for offset, diazo, or photoprint, for example, it will almost certainly be necessary to spend some time in "spotting," that is, in hand-opaquing specks and imperfections in the negative. This is all the more disheartening if the original line material is in good condition. If the original is in poor condition, the effects of dust only add considerably to difficulties in reproduction and to manipulation of the final piece of work in order to obtain some semblance of adequacy.

You can save much time, money, and effort by employing certain preventive steps prior to reproduction. First, however, you must understand the causes of dust problems.

Dust is present in the reproduction laboratory in many forms. These include particles on the piece of line material for reproduction; dust and dirt on the glass of the camera's copy frame or the vacuum frame; dust on the lens; particles suspended in the camera's bellows; dust particles in film holders or on the camera's back; particles suspended in the air; lint or foreign matter contained by paper separators in the film box.

The following are precautions to minimize the effects of dust:

1. Remove particles from the piece of line material for reproduction by

cleaning its surface with a soft, clean brush. A draftsman's cleanup brush is ideal.

2. Clean all glass thoroughly with a suitable glass cleaner that leaves no residue, such as Nu-Arc Cleaner. Nu-Arc is available in spray cans for ease of application.

3. Clean the lens carefully. Use a fine static-free lens brush or lens cleaning paper treated with silicone. (Never, however, use silicone-treated materials for cleaning coated lenses.)

4. Periodically clean out the camera's bellows with a vacuum cleaner. Never use a brush, since dust will merely be shifted from one spot to another.

5. Clean out film holders with a vacuum cleaner or, if available, an air hose.

6. Keep your camera area clean. Check the ventilating system for dirt and dust. A few folds of cheesecloth placed over a system's air supply vents serve as a good filter. If possible, never operate an electric fan in the camera area, as it will only stir up dust. If you use carbon-arc lamp reflectors, clean the fine powder from them periodically. A central-air filtration or air-conditioning system helps to keep dust and dirt at a minimum in the laboratory.

7. If you use interleaved film, brush it off with a fine lens or antistatic brush before inserting it into the camera, or hold it by a corner and shake gently.

Another method to employ that minimizes damage caused by dust is to control exposure and development. When using high-contrast process film to make finished copy, decrease exposure by about 15 per cent and increase development, if possible. This method raises image contrast, thus reducing the presence of pinholes caused by dust. It is recommended only for line work, since exposures necessary for reproduction of halftone and continuous-tone subjects are extremely critical.

If working in areas where dust cannot be controlled by the methods just described, one last-ditch technique you can try is the wetting-agent method. Before developing an exposed piece of film, immerse it in a tray of water containing a wetting agent. Ordinary glycerin can also be used. Leave the film in the wetting-agent wash for a few seconds. The action of the wetting agent breaks the surface tension of the water on the film's surface and rinses off dust particles. After developing, notice that pinholes resulting from dust settling on the film are greatly reduced.

2 | INTRODUCTION TO THE HALFTONE PROCESS

WHEN YOU LOOK at a picture in a book, magazine, newspaper, or other publication, you see various shades of gray tones, as well as dark and light tones (Fig. 2.1). Do you realize that what you are visualizing is not really there?

Examine a picture (commonly referred to as a *halftone*) printed in a publication under a magnifier. Notice that the tones which appear light, gray, and dark are actually composed of dots (Fig. 2.2). These dots are formed in the production of a halftone. In making the halftone negative from the original copy, the technician uses a screen that breaks up the image tones of the photograph or artwork being reproduced into a dot pattern on the halftone negative. This dot pattern is so small that when your eye views the printed halftone, it merges the separate dots and records them as continuous gray tones.

FIG. 2.1 A printed halftone. Note the various shades of gray tones as well as the dark and light tones.

FIG. 2.2 A halftone viewed under a magnifier. Note the various size dots that go into making of the tones.

The sizes of these dots vary in relation to their surrounding areas, but they all appear as individual, solid entities on the negative. To make the reproduction in a publication produced by offset printing, for example, the technician exposes the halftone negative onto a sensitized plate. The plate is then placed on a press, and each dot is printed as a separate, solid entity.

One cannot reproduce an original photograph or piece of artwork directly by the various printing methods without first making a halftone negative from the original. If this were attempted, the result would be a mass of solid ink on a page, since a printing press prints only in solid colors. It does not print in tones. The intermediate halftone process, then, which breaks up tones of the original into solid dot entities, is the key to graphic reproduction of photographs and artwork.

In our discussion of halftones, the meanings of three terms should be understood. These terms are halftone, continuous-tone original, and line negative.

A *halftone* is a reproduction of a *continuous-tone original*. A *continuous-tone* original is a photograph or piece of artwork that has gradations of tones from black to white. A photograph that contains light tones, gray tones, and dark tones, for example, is a continuous-tone original. (Chapter 5 discusses methods of reproducing continuous-tone originals, other than halftone reproduction.)

A *line negative* is one in which there are no tones. When printed, it produces a black image on a white background or, in the case of a reversal, a white image on a black background, with no gray tones. (Chapter 1 discusses reproduction of line material, with emphasis on the line negative.)

Since it is composed only of black and clear dots and areas, a halftone negative is actually a form of line negative. Halftone negatives are made on lithographic film, which is the film used in line reproduction. Lithographic film produces solid black and solid clear portions, with no tonal gradations.

What, then, gives a halftone reproduced in a publication the appearance of having gradations from light to dark? It is simply the size of the dots and their relationship to the amount of background area. If the dots on the negative are small and clear, with much dark area separating them, the image will appear as a light tone when printed. As the dots become larger and the space separating them is reduced, the image appears as various shades of gray tones. If the dots on the negative are small and black, with much clear area separating them, the image appears as a dark tone when printed.

Halftone production is complicated and surrounded by controversy. Many volumes have been published on the subject, and just as many differ as to method and approach. A complete discussion of halftones, therefore, would be impossible in this book. Our purpose is to introduce you to the halftone process and discuss the basic information with which a reproduction technician, or one interested in reproduction, should be familiar.

Unlike other methods discussed in this book, there is no clear-cut and simple formula you can apply to halftone production. There are many ways in which one can produce a halftone. The one selected depends on the experience of the person making the halftone and the printing press to be used in graphic reproduction of the halftone.

A simple offset press, for example, may require one type of halftone production method, while a press of more precision may require another type. Once you become experienced in halftone production and learn the characteristics of the presses at your disposal, you will be able to fashion your method of making halftones to meet press requirements.

In the following sections, two methods of producing halftones—the Kodalith Autoscreen film method and the contact-screen method—are discussed in detail. These are of most concern to the reproduction technician engaged in printing operations.

A third method, the glass-screen method, is introduced. Most reproduction facilities, except for lithographic laboratories, would have little use for this process, since its employment would mean an immediate outlay of money for expensive equipment.

The electronic scanning machines used by photoengravers in halftone production are not discussed, since electronic scanning is extremely specialized and outside the scope of reproduction laboratories in the sense that the term *reproduction* is used in this book.

Before discussing various halftone production methods, Sec. 2.1 will explain certain basic information relative to halftones. An understanding of this section is necessary to avoid confusion when we get into the discussion of production methods.

2.1 Basic Principles

When dealing with halftones, it is important to an understanding of the process to keep certain terms and photographic principles clearly in mind.

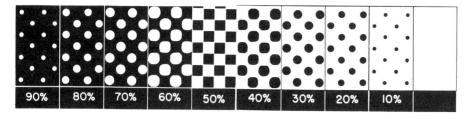

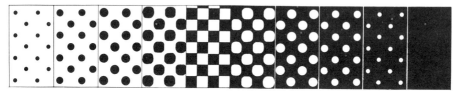

FIG. 2.3   Halftone dot structure. From 10 to 50 per cent dot structures, the clear dot occupies the amount of space referred to. From 60 to 90 per cent dot structures, the percentage refers to the amount of space taken up by the black area.

The terms are *areas* and *dots;* the principles concern *negative* and *positive.*

As has already been mentioned, in the reproducing of a continuous-tone original as a halftone on lithographic film, the image of that original is reproduced as a line negative in the form of black dots on a clear background or clear dots on a black background. This is true for the entire image of the original, except for those portions which are pure black and pure white in nature.

Take, for example, a dark tone of the continuous-tone original's image that is reproduced on the lithographic film. This tone shows up on the negative as little black dots (called *shadow dots*) surrounded by a clear area. On the final print, which is in positive form (the reverse of negative), the clear area of the negative reproduces as dark, while the small black dots reproduce as light, thereby producing an image that is not totally black.

Just the reverse is true in the reproduction of a light image tone of a continuous-tone original. On the negative, this tone appears as small clear dots (called *highlight dots*) surrounded by a black area. When reproduced as a positive, the black area reproduces white, but is broken into a light shade because of the widely spaced dots, which reproduce as black.

When reproducing middle (gray) tones of the continuous-tone image, the dots and the areas surrounding them are almost the same size or they are the same size. This gives the appearance on the reproduced positive of various shades of gray.

Later in our discussion, we shall refer to negative dot structures in terms of percentages, that is, from 10 per cent (shadow) dot structures to 90 per cent (highlight) dot structures. These are terms commonly used in the lithographic industry.

Taken at face value, a dot structure of *x* per cent should mean that a certain percentage of a particular tonal portion is occupied by dots. This is true when speaking of a 10, 20, 30, 40, or 50 per cent dot structure. It is *not* true, however, when referring to a 60, 70, 80, or 90 per cent dot structure (Fig. 2.3).

From a 10 to a 50 per cent dot structure, the black dot on a halftone negative actually occupies the amount referred to in relation to its surrounding area, which is clear.

From a 60 to a 90 per cent dot structure, the percentage refers to the amount of space taken up by the black area, and not the clear dots. In other words, a 90 per cent dot structure on the negative, for example, means that 90 per cent of the area is occupied by black, while only 10 per cent of that portion is occupied by clear dots.

The character of the dots and the surrounding area on the halftone negative changes once a 60 per cent structure is reached. At 60 per cent and thereafter, the dots on the negative are clear, while the surrounding areas are black.

You should not be confused by these terms in another sense. The term *dot structure* in the reproduction and lithographic industries refers to that structure *as it appears on the halftone negative* and not in the reproduced halftone, which is in positive form.

If you keep these factors in mind, you should have little trouble understanding the discussion that follows and, for that matter, the entire process we know as halftone photography.

Before discussing actual processes, however, let us distinguish between the various types of halftones (and their terminologies) you will encounter.

Halftone negatives for photolithography (offset printing) and photoengraving (letterpress printing) are prepared in the same manner, although the screens employed in each may differ slightly. In making the printing medium for reproduction by most offset methods, the halftone negative is exposed directly onto a sensitized plate, which is developed before being placed on the press.

In other cases of halftone preparation for offset printing, such as deep-etch platemaking, where printing plates are etched to provide a medium for extra long printing runs and maximum fidelity of the image, a film positive is needed rather than a negative. The halftone negative is contact-printed, under vacuum, onto a sheet of lithographic film or lithographic strip-off film to produce the positive. Since the halftone dot structure must be reproduced faithfully on the film positive, care should be taken in the contact process.

In letterpress printing, the halftone negative is exposed onto sensitized metal, the background of which is etched to provide a raised image. This metal plate is put on a block, and the block is locked into a form for printing.

There are many varieties of halftones. Here, we shall mention only the most common types. In photolithography, all manipulations to obtain these variations are made in the preparation of the halftone negative,

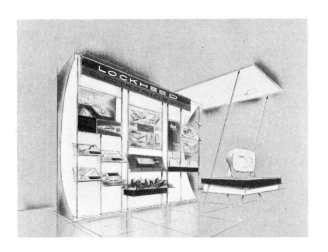

FIG. 2.4   A square halftone. (*Courtesy of Lockheed Electronics Company.*)

whereas in photoengraving some of the manipulating may be performed in the etching of the plate.

The most common types of halftones are as follows.

Square Halftone. A square halftone (Fig. 2.4) is a halftone image having four square corners, with the entire area occupied by a screen pattern. A square halftone does not necessarily have to be square in form. As long as it has square corners, whether it is actually square or rectangular in nature, it is referred to as a square halftone. It is the most common type and is the easiest to produce, since it does not involve masking of any part of the image.

Silhouette Halftone. In a silhouette halftone (Fig. 2.5) some portion of the background is removed by opaquing to produce an image against a pure white background. If the original being copied has a pure white background to start, the silhouette effect can be created by the highlight-bump technique discussed in Sec. 2.3.

Surprint Halftone. A surprint halftone is a square halftone with a black or colored solid image printed directly on top of it. (Fig. 2.6). Surprint halftones make use of both a halftone negative and a line negative. The halftone plate is usually printed first, and the line plate is printed over the halftone to obtain a surprint halftone.

Highlight Halftone (or Dropout Halftone). In a highlight (dropout) halftone, the halftone dots are removed from a portion of the image area to accentuate a highlight (Fig. 2.7). The area to be highlighted is opaqued out in the halftone negative if it is to be reproduced by photolithography or by removing the dots from the negative or printing plate if the halftone is to be reproduced by photoengraving.

The screens used in the making of halftones are identified by the num-

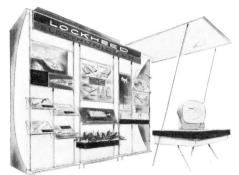

FIG. 2.5   A silhouette halftone. (*Courtesy of Lockheed Electronics Company.*)

FIG. 2.6 A surprint halftone. (*Courtesy of Lockheed Electronics Company.*)

ber of ruled lines on them per inch. Thus, a 133-line screen produces 133 lines per in. on the halftone negative.

The screen you use, whether 65-line, or 133-line, etc., depends on the type of press and paper you employ in printing and, of course, on the result desired. Many printing presses have limitations that dictate the screen rulings you can use. Precision presses for offset and letterpress printing can usually accommodate the finest screens, while the smaller, less versatile machines cannot.

Papers for printing also play a role in the type of screen selected. Newspapers, for example, generally use a coarse 50- or 60-line screen because of the high ink absorption rate of newsprint, the rough surface of newsprint, and the method used to print the image, which is usually rotary letterpress.

Very fine screens (133-line and up) produce halftones of higher quality than the coarser screens (120-line and below). When extremely fine screens are used, the printed halftone is often mistaken for a photograph, since the dots are so close together, thus making them harder to see. Generally, fine screens reproduce detail better than coarse screens. Consequently, they are employed when detail is essential.

Coarse screens are widely used in the preparation of halftone prints (called velox prints). The halftone negative is made with a coarse screen, usually one that is two times larger than the final size will be. The halftone negative is then contact-printed, under vacuum, on a sheet of glossy photographic paper. This print, possessing all the dots of the negative, can now be used in a mechanical layout or by itself for making future halftones.

FIG. 2.7 A highlight halftone.

For example, if the halftone negative were made double size with a 65-line screen—let us say the image size is 4 in., giving a halftone negative size of 8 in.—a 50 per cent reduction would bring the image down to the correct size and give the equivalent of a 130-line screen.

2.2 Kodalith Autoscreen Film Method

For the reproduction technician only generally familiar with lithographic techniques, the Kodalith Autoscreen film method offers the opportunity to produce acceptable halftone negatives with relatively little training. The Autoscreen technique can be implemented at less cost than other halftone methods, since the film is employable with cameras found in most reproduction laboratories. The only expense needed to establish a halftone capability in your laboratory is the cost for film and chemicals.

Kodalith Autoscreen film is a high-contrast orthochromatic material, with a built-in, pre-exposed 133-line screen. Since the screen is pre-exposed and built right into the Autoscreen material, there is no need to employ a separate mechanical screen to form the dot pattern on the film. Most other methods of halftone production, however, require that a separate screen be placed between continuous-tone original and film.

Unlike other photographic materials, Autoscreen film is not equally sensitive to light throughout its emulsion. This emulsion is composed of thousands of individual segments, with each segment having its own sensitivity, separate and distinct from the others. Each of the individual segments is extremely sensitive to light at its center, but the emulsion decreases in sensitivity as it gets further and further from this center point.

Despite the complexity of the Autoscreen composition, it produces results similar to other types of halftones made on lithographic films. When a continuous-tone original is copied, light hitting the dark (shadow) areas of the original reflect only a minimum amount of light to the film. This results in very weak exposures to the film, producing an image only in the most sensitive portions of the emulsion, that is, in the center of those sensitive segments exposed to light.

This shadow exposure produces small black dots, which are surrounded by a large-size clear area, on the halftone negative. When exposed on a plate, the clear area of the negative reproduces dark, but not black, since the small, widely spaced black dots of the negative, which reproduce white, introduce white into the reproduced tone. A maximum amount of ink from the press is picked up by the plate, and the tone is printed as dark.

A greater amount of light is reflected from the original's middle (gray) tones to the Autoscreen emulsion. These middle tones expose the less sensitive portions of the film, as well as the highly sensitive center portion, producing larger dots that are almost the same or are the same size as the area surrounding them. This type of dot and area structure picks up less ink from the press and prints as various shades of gray tone.

The light tones (highlights) of a continuous-tone original present the maximum in reflectiveness. These highlights are recorded throughout the

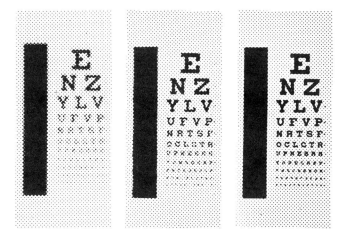

FIG. 2.8 Reproduction by halftone methods. The type matter was reproduced by glass crossline screen (*left*), Kodak magenta contact screen (*center*), and Kodalith Autoscreen ortho film (*right*). (*Courtesy of Eastman Kodak Company.*)

entire surface of the individual film segments to which they are exposed, producing small clear dots and large-size black areas on the halftone negative. On the press, these dots pick up a minimum amount of ink and have the greatest amount of white space around them, thus printing as light tones.

Autoscreen film requires shorter exposure times than other films used in halftone work. Since the film does not employ a separate mechanical screen, exposure times are reduced.

Since Autoscreen is a fine detail material, you can produce excellent halftones from continuous-tone originals possessing text or lettering. This is done by exposing text or lettering, which can be in the form of an acetate overlay placed over the image area or which can be directly on the image area, at the same time you expose the picture. This is done on one negative. The text reproduces legibly (Fig. 2.8).

When employing methods other than Autoscreen, it is often necessary to photograph text and picture separately to obtain separate negatives of equal legibility. These negatives are then stripped into a single printing form, called a goldenrod. A printing plate of the combination text and picture is made and both are printed as one. (For a discussion of stripping, see Sec. 2.5.) If two-color printing is employed, the separate negative would have to be reproduced on two printing plates, with each printed individually.

Autoscreen film provides greater highlight contrast, but lower shadow contrast, than the contact- and glass-screen methods of producing halftones. Excellent halftones are produced from continuous-tone originals containing sharp highlight contrast, thus fulfilling the major requirement in halftone production of distinct highlight separation (Fig. 2.9). In those cases where shadow detail is of major concern, however, halftone methods employing a separate mechanical screen may be more suitable.

Being orthochromatic, Autoscreen can be handled safely in the darkroom under a red safelight. A Wratten Series 1A filter (light red) in a safelight housing containing a 15-watt bulb is ideal. Keep the safelight *no less* than 6 ft from the film.

This is important! If Autoscreen is handled less than 6 ft from the safelight for more than a minute, there is danger of fogging the material and impairing the quality of the dot structure.

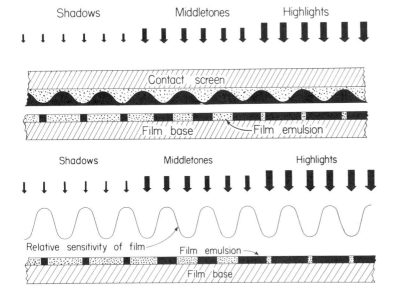

FIG. 2.9 Reproduction by Autoscreen. Though the same dot-patterned effect is achieved on Autoscreen ortho film as with other halftone reproduction methods, there is no screen through which the light is filtered. Instead, the film's sensitivity varies across the surface of the emulsion and the size of the dots is proportional to the amount of light striking the film. (*Courtesy of Eastman Kodak Company.*)

The following is a step-by-step guide to use in exposing and processing Autoscreen film. One word of caution beforehand, however: to produce halftones of maximum quality, no matter what method is employed, the continuous-tone original being copied should possess a full range of tones and definite highlight and shadow separation. A good, even gloss is desirable, since glossy prints reproduce better than matte prints because of their high reflectiveness.

Exposure. Place a sheet of Autoscreen in a film holder for exposure by view camera or on the vacuum back of a graphic-arts reproduction camera. (Autoscreen is designed to fit into standard-size film holders.) Two exposures are necessary: a *detail exposure* and a contrast-control exposure, called the *flash exposure*. Each is separate from the other and is accomplished in a different manner.

The *detail exposure* is made first in the camera, with the time of the exposure based on the highlight density of the original being copied. Its purpose is as its name implies: to record the details (highlights) of the original. In other words, the detail exposure controls highlights.

A typical detail exposure (with lights arranged at 45° angles to the continuous-tone original) is about 25 sec when exposing with two No. 2 photofloods, and about 16 sec when exposing with two 35-amp carbon-arc lights. Photofloods should be placed 3 ft from the subject, and carbon arcs should be placed 4 ft from the subject. These exposures are based on a 1:1 (same size) reproduction with a lens opening of *f*/22.

The *flash exposure* is made with a yellow light. Do *not* use the camera. Its purpose is to control the film's contrast by evenly fogging the entire halftone area. The flash exposure produces the black dots of the shadow area on the negative, thus bringing contrast more into line and providing a more pleasing effect. It does not affect the highlights.

The average time of a flash exposure is from 20 to 30 sec using a 25-watt bulb in a safelight housing that contains a Wratten OA (yellow) filter. The safelight is usually held 6 ft from the film.

By increasing flashing time, you can actually change the size of the shadow dot structure, thus making the tone on the printed page lighter.

FIG. 2.10 Calibrated gray scale. Use this to determine exposure times for any halftone method. (*Courtesy of E. I. Du Pont de Nemours and Company, Inc.*)

Remember this fact concerning the detail and flash exposures: the detail exposure controls the highlights and the flash exposure controls the shadows. They are separate exposures, and they are independent of each other.

The times of exposures cited above are for average situations. Actually, each reproduction operation has its own peculiar characteristics that may lengthen or shorten exposure time. For this reason, it is best to make test detail and flash exposures until you learn the characteristics of your setup.

Detail exposure tests are made by photographing a continuous-tone gray scale at a same-size (1:1) ratio. A gray scale, which can be purchased from your graphic-arts supplier, contains pure white and pure black tones and progressive shades of gray between these tones (Fig. 2.10). This test should be made to determine the exposure required to produce a 90 per cent highlight dot structure. Once this has been accomplished, the exposure time obtained will produce the proper dot sizes of all other highlight and gray portions of the continuous-tone originals you copy. If not, reduce or increase the detail exposure, as required.

Begin your detail exposure tests at the times recommended above, that is, 25 sec when exposing with two No. 2 photofloods and 16 sec when exposing with two 35-amp carbon-arc lamps. Regardless of the light system employed, exposures should be made at a lens opening of $f/22$.

If the relative size of the 90 per cent dots is not approximately the same as the ones illustrated in Fig. 2.11, make other tests, increasing exposures at 5-sec intervals until dots of the approximate size illustrated are obtained. This time is the one at which your best detail exposure will be produced.

After determining the detail exposure time for a 1:1 reproduction, it is advisable to make up a chart listing the proper detail exposure times for various size reductions, such as 50 and 75 per cent, for the types of originals you may be reproducing. For example, a light original containing little contrast requires a different detail exposure than dark or high-contrast originals at different reduction ratios.

Once the correct detail exposure time is obtained for your setup, you must now find the flash-exposure time that produces the maximum in

FIG. 2.11 Dot structures produced by Autoscreen film. On the left is the 90 per cent highlight dot; on the right is the 20 per cent shadow dot. Magnification: eight times.

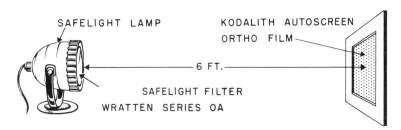

SAFELIGHT LAMP

KODALITH AUTOSCREEN
ORTHO FILM

6 FT.

SAFELIGHT FILTER
WRATTEN SERIES OA

FIG. 2.12   One method of flashing Autoscreen. The flashing lamp is suspended from the ceiling, and the uncovered film is placed flat on a workbench 6 ft below. (*Courtesy of Eastman Kodak Company.*)

shadow registration (Fig. 2.12). When using separate mechanical screens to produce halftones, the shadow dot structure you wish to simulate in this test is one of 10 per cent. However, because of the decreased shadow quality reproduction encountered when using Autoscreen, a shadow dot structure of 20 per cent is the one you should strive for in this test and in all your flash exposures (Fig. 2.11).

To perform a flash-exposure test, thus determining the correct time of flashing for your setup, proceed as follows:

1. Position an unexposed sheet of Autoscreen film in front of the yellow light. Do not expose this sheet to the gray scale that you use when testing for the detail exposure.

2. Cover all but a small strip of the film with a sheet of black paper. Expose this strip to the flashing light for 5 sec. Move the black covering paper down, leaving more of a strip visible, and expose this strip and the one previously exposed for another 5 sec. Repeat this operation all the way down the sheet of film (Fig. 2.13).

The result is a test sheet with strips of film exposed for a minimum of 5 sec and a maximum for the number of seconds you exposed the first strip. For example, if you exposed nine strips of the film, the last strip you exposed will have received a 5-sec flash, the first strip you exposed will have received a 45-sec flash, and the strips between these two will have been exposed from 10 to 40 sec at 5-sec intervals.

3. Develop the test sheet as recommended below.

4. Check the results on a light table, comparing Fig. 2.11 with the dot structure produced on the test strips. The strip that compares most closely with a 20 per cent shadow dot is the one that is correct, and the time needed to obtain this dot structure is the one to use for flashing.

Thus far we have determined by two separate test exposure methods what the correct highlight and flash-exposure times should be for a 1:1 reproduction. Now, integrate the two exposures to make a final halftone negative. As you gain experience, you will be able to make halftone negatives without first making test exposures by looking at the highlight-density ranges of a continuous-tone original and judging proper exposure time.

Remember this important point after determining the correct detail and flash-exposure times for a same-size (1:1) reproduction: as with other types of reproduction work, the detail exposure is affected by a reduction in the size of the reproduction. For example, a 50 per cent reduction changes the time of your detail exposure. But it does *not* change the time of the flash exposure. The only factor affecting this time is the density of the original's shadow areas.

Processing. Autoscreen film should be developed in Kodalith developers A and B, which come in powder form. Kodalith developer is also available in liquid form, but it is *not* recommended for developing Autoscreen. If you employ a lithographic developer other than Kodalith, make sure the development times are exactly the same as that recommended for Kodalith, since this time (3 min) is critical and must be exact.

Prepare Kodalith solution before making the exposures, so that the solution can cool down to the proper developing temperature (68 to 70°F). To prepare the solution, mix 1 part of Kodalith A and 1 part of Kodalith B in 1 part of water.

Two minutes of the three-minute development time is spent in vigorous agitation of the film in the Kodalith. During the remaining minute, the fine-line development technique is employed. The film is allowed to remain perfectly still in the developer.

To agitate the film, rock the developing tray from corner to corner, so the solution makes a complete movement around the tray. After exactly 2 min, stop agitation, and let the film lie perfectly still in the tray. The film should be pressed down to the bottom of the tray, so that it does not float around in the solution.

Only one halftone negative should be developed at a time, and only enough developer to cover the film should be used. If there is too much developer in the tray, perfectly still development is not possible. Any movement during the still-development phase of processing increases the danger of producing streaks on the film.

These streaks show up in printing. One precaution is to use the smallest-size tray possible for the film with which you are working. For example,

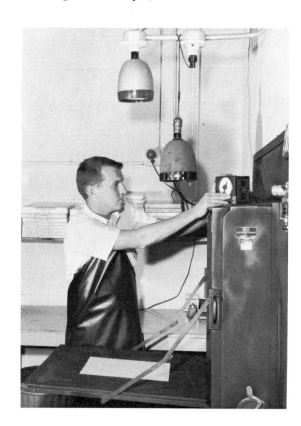

FIG. 2.13 A second method of flashing Autoscreen. If a reproduction camera is used, drop the vacuum back after making the detail exposure. The flash is made with a safelight, equipped with on OA filter, suspended from above and connected to an interval timer. It is not necessary to build up any vacuum, since Autoscreen film has a built-in and pre-exposed screen and not a separate mechanical screen which requires good contact between film and screen. (*Courtesy of Lockheed Electronics Company.*)

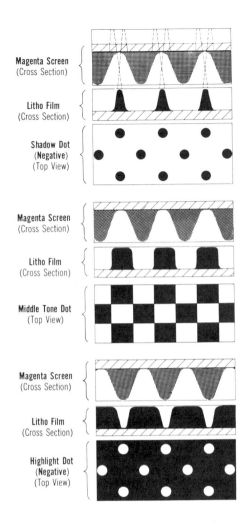

Magenta Screen
(Cross Section)

Litho Film
(Cross Section)

Shadow Dot
(Negative)
(Top View)

Magenta Screen
(Cross Section)

Litho Film
(Cross Section)

Middle Tone Dot
(Top View)

Magenta Screen
(Cross Section)

Litho Film
(Cross Section)

Highlight Dot
(Negative)
(Top View)

FIG. 2.14 Effects of a magenta screen. This cross section illustrates how magenta screen reacts on lithographic film to produce shadow, middle tone, and highlight dot effects. The gray screen works in a similar manner. (*Courtesy of E. I. Du Pont de Nemours and Company, Inc.*)

when developing an 8- by 10-in. sheet of film, use an 8- by 10-in. tray, and not an 11- by 14-in. tray.

At the end of 3 min, remove the negative from the developer immediately, and immerse it in a stop bath. After 15 to 20 sec in the stop bath, transfer the negative to a fixing solution. The stop bath, fixing, washing, and drying phases of processing are accomplished in the normal photographic manner.

One final caution: Whenever you handle Autoscreen film, be very careful not to scratch the negative. You cannot opaque a halftone negative, except in those areas where special effects, such as highlighting, are desired.

2.3 Contact-screen Method

The contact screen is probably the most widely used of all the methods for producing halftones. There are two types of contact screens: the magenta screen and the gray screen.

The magenta screen is used when making halftone negatives on orthochromatic lithographic film. The gray screen is employed for making color separation halftone negatives on panchromatic lithographic film and conventional halftone negatives on orthochromatic lithographic film. Of the two, magenta screens are more widely used in the production of conventional halftones.

Both magenta and gray screens are employed the same way when used in the reproduction to halftone form of black-and-white continuous-tone originals, and the procedures outlined in this section apply to both screens, except as noted.

Contact screens are placed in direct contact with the film on a vacuum back of a graphic-arts reproduction camera. Since the method entails a separate mechanical screen that must be in direct contact with the film, only those reproduction facilities possessing a graphic-arts reproduction camera with a vacuum back can employ the contact-screen method.

By the contact-screen method, one can produce high resolution halftone images. The image is formed by light passing through the vignetted dot pattern of the screen and acting on the film's emulsion (Figs. 2.14 and 2.15). Contact screens are available for both photoengraving and lithography.

In addition to producing high-resolution images, the contact-screen method offers other advantages. It is a comparatively inexpensive process when compared to a method such as glass screen. It is flexible and simple in usage. However, one caution should be stressed immediately. Contact screens are usually made of acetate or triacetate, and are fragile. They must be handled with as much care as you handle the film with which they are employed to prevent cracks, kinks, fingermarks, scratches, and moisture marks.

Magenta contact screens are designed primarily for illumination by carbon-arc lamps. If you use photoflood lights, you must place a Wratten Series No. 80 filter over the camera's lens. The magenta dye on the screen transmits bands of light that are designed for lamps having a specific color temperature. Carbon arcs meet this requirement. With sources other than carbon arcs (photofloods, for example), you must change the color temperature of the light to simulate that of arc lamps.

Gray contact screens can be used equally as well with either carbon arcs or photofloods, without a compensating filter.

In making a halftone negative with a contact screen, you should first

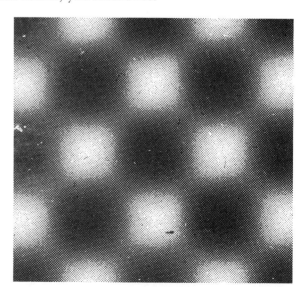

FIG. 2.15 Dot structure produced by a magenta screen. Notice the vignetted edges of the dot structure.

determine the detail exposure. You are interested in obtaining the proper exposure time to produce a correct 90 per cent highlight dot. When printed, the highlight dot forms the very small black dots that break a white area into a tone. If the highlight dot on the negative is too large, the final print will contain darker highlights than desired. They will look something like an underdeveloped, overexposed photographic print.

To determine the correct detail exposure, proceed as follows:

1. Place a continuous-tone gray scale in the easel of your camera. For accurate testing, the gray scale should have a greater or the same density range as the originals you will be reproducing. The Du Pont calibrated gray scale is an excellent medium, as is the density guide gray scale contained in the Kodak graphic-arts computer package.

Set the camera's scale for a same-size (1:1) reproduction. After you have determined the basic exposure, you will be able to calculate exposures for reductions in copy by using the 1:1 exposure time as a constant. For example, the approximate exposure time for a 50 per cent reduction is generally one-half the exposure time of a 1:1 copy. However, tests should be conducted to obtain the absolute correct time when making reductions.

2. Place a sheet of orthochromatic lithographic film on the camera's vacuum back, and gently lay the magenta or gray screen over the film, so that the emulsion of the contact screen and the emulsion of the film are in contact with each other. Turn on the vacuum motor, and ensure that good contact is made.

The contact screen you use should be slightly larger than the film, resulting in a better contact between screen and film. To make the contact even more firm, press film and screen together by use of a soft rubber roller or a dry photo chamois. (By the way, when removing dust from the screen, use a fine camel's-hair brush *carefully*.)

3. Make a series of exposures to determine the correct one for a proper 90 per cent highlight dot structure in the white area of the mounted gray scale. Two carbon arcs should be used at 45° angles to the easel, 4 ft from the subject. Start your exposures at 25 sec, with the lens stopped down to $f/16$. Increase the exposure for each recurring test by 5 sec until the last test is accomplished at 45 sec. Mark each negative with its respective exposure time.

4. Develop test films in a freshly mixed lithographic developer at the proper temperature (usually 68°F). Developing times may vary, depending on the developer used; so follow package instructions carefully. Develop each sheet individually. Then, immerse in stop bath and fix. Films should be agitated in the developer in the normal manner.

5. After negatives are developed, check each one on a light table with a good 10× or 12× lithographer's magnifier against the 90 per cent dot structure portrayed in Fig. 2.11. After you have ascertained the correct exposure, place a good glossy photograph in the easel, and make a test shot at this exposure. After developing the negative, you can determine if this exposure produces the desired results. If not, you can make minor adjustments in exposure time, decreasing exposure time slightly if the clear dots are too small (highlights are too light), or increasing this time slightly if the clear dots are too large (highlights are too gray).

As with the Autoscreen method of producing halftone negatives, the detail exposure, which controls the highlights, must be followed by a flash exposure, which controls the shadows. When using the contact-screen method, the shadow areas should have a proper 10 per cent dot formation to produce good halftones for printing.

The flash-exposure procedure with contact screen is accomplished in the same manner as with Autoscreen (see Sec. 2.2), but there are some differences in equipment and method, as follows:

1. The exposure is made on the vacuum back of the camera, with the screen in place. The back is opened, since this exposure is *not* made through the lens. Maintain the correct vacuum after making the detail exposure (Fig. 2.16).

2. The equipment needed for flashing with a magenta contact screen is a 7½ watt frosted bulb in a safelight equipped with a No. 00 (yellow) filter. The safelight should be held about 4 ft from the screen and film, and will cover an area of about 16 by 20 in.

3. The average time of flash is from 15 to 25 sec, but a test should be conducted to arrive at the exact time. Tests are made on unexposed film (orthochromatic lithographic) and accomplished in the same manner as with Autoscreen, with a variation only as to length of time. Begin exposure at 3 sec and progress by 3-sec intervals until the entire sheet of film has been exposed. A minimum of nine exposures should be made, so that the first strip exposed receives a flash of at least 27 sec.

There are certain variations to the normal manner of using contact screens that can be employed to make better halftones. One of these, the *highlight bump,* permits the making of good halftones from continuous-tone originals not possessing a full range of tones. The method permits you to increase highlight contrast when reproducing a flat or extremely high-key original.

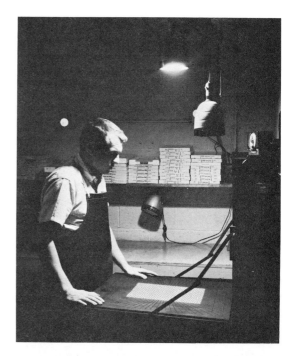

FIG. 2.16   Flash exposure with contact screen. The filter is an 00. The film is flashed with the screen in position on the vacuum back of the reproduction camera and the vacuum turned on. (*Courtesy of Lockheed Electronics Company.*)

A high-key original is one which possesses light and middle tones (light grays), but very little or no black or dark tones. High-key originals can be considered as being low in contrast, since there is little or no noticeable contrast between black and white, or between dark and light. They are *not* considered as "flat" originals, since "flat" refers to an original that has been exposed or developed improperly, giving an overall gray appearance.

The highlight-bump technique requires an additional short exposure of the film to the original after the normal detail and flash exposures have been made. This short exposure is made *without* the contact screen.

After completing the flash exposure, gently lift the screen from the film, leaving the film on the camera's vacuum back (at proper vacuum). Close the camera back and re-expose the film to the original for a short time. This so-called *highlight bump* is likely to be 5 per cent of the detail exposure for an original of limited density range and less than 5 per cent for an original of more limited density. You can determine density range by comparing the original to the gray scale.

Variations in exposure must be determined by trial and error. No effective guide can be given, since shooting and setup conditions vary from laboratory to laboratory. Once you have determined the proper exposure variation for making a highlight bump with the type of originals you work with, however, this variation can be used most of the time. Usually, exposure differences for highlight bump vary only slightly from the detail exposure. The highlight-bump exposure affects only the highlights by decreasing the size of the clear dot.

A technician is often called upon to reproduce an original without background in the final halftone reproduction. This type of rendition is referred to as a *blockout* or *dropout*. It is made by simply increasing the highlight-bump exposure to the point where the background is absolutely black on the negative or at least to the point where it is too dense to print through. With this method, unfortunately, there is a slight loss of image quality, but this is more or less compensated for by a definite increase in image brilliance caused by the increased highlights. To employ the blockout method, the original to be copied should have a white background. If it does not, the background can be opaqued out on the halftone negative.

The blockout technique is frequently used when combinations of line and tone are to be reproduced as halftone negatives for printing, since it eliminates the need for making double negative exposures in the plate-making stage; that is, one negative for the line copy and one for the tone copy.

There are certain precautions to follow when making halftone negatives. These are as follows:

1. Beware of reflections from surrounding area and objects. Reflections, which could ruin an otherwise good reproduction, can easily be eliminated by painting the camera-room walls and ceiling with a flat-black paint (Fig. 2.17). Keep overhead lights turned off when making exposures.

2. Beware of lens flare. Use a coated lens and a black easel to guard against this condition.

3. Beware of safelight fog. Check safelights for white-light leaks. Make sure you use proper filters and bulbs.

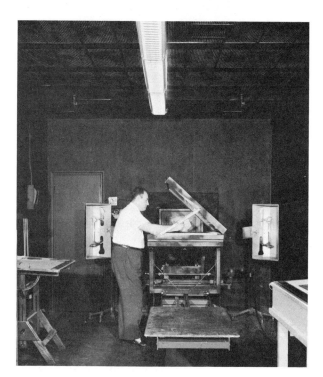

FIG. 2.17 Method of reducing reflections. A camera area painted with a flat black paint will reduce reflections and glare that could ruin halftone negatives. (*Courtesy of Lockheed Electronics Company.*)

4. Beware of ruinous conditions caused by faulty developing. Developers should be used at their recommended temperature (usually 68 to 70°F) to achieve maximum and consistent results. Always use a fresh developer and never let the developer set too long in an open tray. Lithographic developers oxidize rapidly after the A and B solutions are mixed.

2.4 Glass-screen Method

The glass-screen method of producing halftones is an extremely specialized one. Only a general familiarization with this method is included here, since few reproduction laboratories would have occasion to use it.

The glass screen consists of two sheets of glass cemented together. Each sheet has fine parallel lines ruled on it, with these lines running at right angles to each other, forming a window-screen pattern (Fig. 2.18). The width of the lines usually equals the width of the openings between them.

Glass screens are available in from 45 lines per in. up to 400 lines per in. The latter is used for specialized work.

FIG. 2.18   Enlarged view of rulings on a glass screen.

The screen is used between a camera's lens and the film plane at a predetermined distance from the film, depending on the screen's ruling. It breaks up the image of the continuous-tone original being projected on it through the camera's lens into a series of small squares (dots). In effect, each tiny opening on the screen acts much the same as a pinhole lens and photographs only a tiny part of the copy. Thus, the continuous-tone image is reduced to a series of dots of equal density, but of different size, depending on the density areas of the original.

Glass screens have long been regarded as the finest mediums for producing high-quality halftones, but there is a difference of opinion as to their specific handling. They are used mainly by the lithographic and photoengraving industries. The high cost of the screens and the cameras required in their use, plus the extremely high level of experience needed to work with them, make them especially applicable to professional lithographic and photoengraving operations.

2.5 Basic Stripping Techniques

After the halftone negative has been made and before it is given to the printer, there is an intermediate step known as *stripping*. It is applicable mainly to offset printing operations.

Complex stripping operations are usually accomplished by an experienced layout man, who is often an artist. However, in emergencies, they could be done by the printer or the reproduction technician. For this reason, it is important for these people to understand the principles of basic stripping.

Stripping is the preparation of negatives (or positives) made in the camera room for platemaking. The one doing the stripping performs the same function for the offset printing process as the makeup and lockup men perform in letterpress printing.

The person doing the stripping usually works from an original layout that tells him exactly where each element is to appear in the final printing. In offset work, these elements are arranged on a layout sheet called a *printer's flat.*

The printer's flat permits the negative (or positives) to be positioned in their proper page location before platemaking. These layout sheets come in many sizes to meet various types of press requirements and in many colors. The most commonly used color is yellow or orange. This color is employed since it holds back the ultraviolet light used in exposing the finished plates, while permitting the desired cutout areas to print through.

Because of its color, this type of printer's flat is called the *goldenrod sheet.* It is available in either ruled or unruled form. The unruled variety is ordinarily used by the experienced stripper, who wishes to make his own layout. For normal stripping operations, however, the ruled goldenrod sheet is the most practical. Ruled lines permit the positioning of negatives (or positives) in the exact spot and provides a quick check to ensure proper alignment of the image. (By proper alignment is meant the correct positioning of the image so that it is not crooked on the page.)

The person doing the stripping works over the light of a light table. This allows him to see through the negatives (or positives) with which he is working and provides him with the light whereby he can position correctly the negatives (or positives) on the goldenrod sheet.

The materials and tools normally needed for stripping are as follows:

1. Visual tools. A light table or light box large enough to accommodate the maximum size printer's flat that will be used.

2. Tools for cutting. Razor blades (safety edge), a metal straightedge, metal triangles, and scissors.

3. Stripping materials. Goldenrod sheets, transparent cellophane tape, and red or black acetate tape (the "see-through" type is preferred since it holds back light, but permits one to see what has been covered). These materials are available from graphic-arts suppliers.

4. Touch-up materials. Opaquing fluid (quick-dry opaque is the best) and opaquing brushes. These are used to touch up undesirable spots on the negative. They, too, are available from graphic-arts suppliers.

5. Optional material. Other useful items to hasten the work are double-sided tape and a double-sided cellophane-tape dispenser, available from graphic-art supply dealers.

The basic type of stripping job is encountered with a mechanical layout that is laid out in the desired position and contains centering marks. The layout is shot by the reproduction technician, and the final negative shows these centering marks, making positioning during stripping simple.

The negative is placed flat against the light table, so that it is right-reading. The goldenrod sheet (ruled) is placed on top of the negative, and the center lines of the goldenrod sheet are positioned over the center marks on the negative.

One corner of the goldenrod sheet is carefully lifted, and a small piece of double-sided cellophane tape is placed on the negative. The goldenrod is then pressed down on the negative, joining the two corners together. This process is then repeated with the opposite corner, resulting in two corners of the goldenrod joined firmly with two corners of the negative.

The joined goldenrod negative is then turned over, and tape is applied all around the edges to secure negative and goldenrod together. (When making the joint at the corner, be certain that the corners of the golden-rod and negative are secured tightly, so that one or the other does not slip when you turn the two over.)

After securing goldenrod and negative together, you must cut out the goldenrod to permit the printing areas to show through. This is easily accomplished by using a sharp safety razor. Simply flip the joined goldenrod negative so that the goldenrod is on top. Then, carefully cut out a window in the goldenrod in the desired area. Be careful not to cut the negative. When the image shows through, you can now touch up any minor pinholes in the negative with opaquing fluid, completing the operation.

An alternate method of accomplishing this type of stripping is to do it in reverse. Position the goldenrod sheet on the light table with the ruled side facing down. Place the negative on top of the goldenrod with the emulsion facing you, that is, with the reverse reading image facing up. Position the center lines of the goldenrod with the center marks on the negative by looking through the negative. Then tape together as described above and make the cut out.

When taping, use only a limited amount of tape on the layout, and never use tape heavier than the acetate variety. The goldenrod-negative sandwich, taped together, is presented to the printer for platemaking. If the tape is too thick, such as the kind of thickness presented by conventional masking tape, it may cause troubles in making of the plate. Contact between sensitized plate and the negative must be as absolute as possible during exposure. This contact could be hindered by the thickness of the tape holding the goldenrod and negative together.

3 | CAMERA REPRODUCTION OF PRINTED-CIRCUIT ORIGINALS

THE SPECTACULAR GROWTH of the electronics industry since World War II has added an innovation called the *printed circuit* to the tasks of many reproduction technicians. A printed circuit is an electric circuit manufactured by the general technique in which the wiring and/or the components are developed by etching away a conductive metal coating from an insulating board. Photographic reproduction is one of the most important steps prior to the actual etching process.

Heretofore a technique engaged in by those industries concerned with production of military electronic equipment, printed circuitry has suddenly made a dynamic impact on industry as a whole. TV sets, radios, computers, and an increasing number of other electrical products are being equipped with printed instead of hand-wired circuits, since the former are less expensive to produce, easier to repair, and more reliable. For the reproduction technician, the art of printed circuitry offers a challenge, because it requires him to use to the fullest his versatility and knowledge.

Printed circuitry is a consequence of World War II, when demands were made on industry to find a quick, inexpensive method of mass producing, with a reduced labor force, the electronics equipment needed for war. Hand-wiring techniques, which had been employed exclusively to this period, required too much time and manpower.

Relying initially upon the findings of Dr. P. Eisler, who founded printed circuitry in 1940, the electronics industry met the challenge of the times and has gone on to perfect printed circuitry to a point where, conceivably, the practice of wiring equipment by hand will be antiquated within a few years.

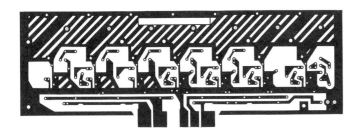

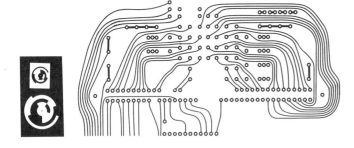

FIG. 3.1 Samples of complex printed-circuit patterns.

This chapter discusses common problems inherent in printed-circuit reproduction. Five areas are outlined: the original drawing (layout), the camera, illumination, the film positive, and selective photoetching. The plating and etching processes required in the production of the actual printed-circuit board, which occur after reproduction, are not discussed, since these operations take place in a plating shop and not a photographic laboratory.

3.1 The Original Drawing (Layout)

An artist or draftsman is usually responsible for laying out the printed-circuit original, although at times an engineer may perform the task.

Originals containing simple circuit patterns can often be drawn to the size specified for the final printed-circuit board. If laid out on a translucent or transparent material, these could be given directly to the plating shop for production of the board, without having to undergo the reproduction process.

Most originals, however, contain intricate pattern detail and configuration (Fig. 3.1). They are usually drawn to a size that is two to twenty times larger than the size stipulated for the final board, and thus require reduction by photographic means. These oversize drawings compensate somewhat for errors in layout, since the larger the layout, the larger the permissible error. In the photographic reduction of the layout to final board size, the error is reduced proportionately. (Error refers to discrepancies in size, ragged lines, and other imperfections in the original layout.)

Accuracy is often a requirement for printed-circuit boards, particularly if both sides of a board have circuit patterns that must register with one another for the purpose of inserting components. High-frequency circuits are especially critical, since an error in the correct length or width of a pattern could cause electrical difficulties in the form of capacitance or inductiveness. In low-frequency circuits, extreme accuracy is not usually as vital, although an effort should be made to achieve it.

Generally, the slightest error in layout or reproduction could result in a faulty board being placed in an electrical system and, consequently, in a system breakdown. Thus, the reproduction technician must exercise caution in all facets of his work. When dimensions are critical, he must have control over every step of the reproduction process. The slightest error, perhaps no larger than $\frac{1}{64}$ in., could be as ruinous and as unacceptable as an error of 1 in. or more (Fig. 3.2).

Since the composition of an original drawing often dictates the method of reproduction, the technician should be familiar with the mediums employed by the draftsman or artist in layout. The five common mediums are tape, ink, scribing, peel-off, and scribe–peel-off.

Tape Method. Printed-circuit patterns are most often laid out with tape, which is placed on a sheet of translucent material. The layout man places the translucent material over the circuit sketch, which is usually drawn in pencil, and lays both on a light table. Because he can see through the translucent material to the sketch below, he is able to follow the pattern lines and place his tape in the correct position.

Tape offers the experienced draftsman certain advantages: it is easy to handle, it permits rapid manipulation, and it is available in all shapes and sizes (either in special dispensers or in small rolls) to form any required pattern without having actually to cut that pattern.

Most tapes are manufactured in a rough textured material that has a surface similar to that of black photographic masking tape. Usual difficulties encountered when photographing tape originals are as follows (see Sec. 3.3 for an explanation of how to overcome these problems):

1. The surface of the tape can cause reflections.

2. Because the tape has a definite thickness, slight shadows and reflections could be created around its edges. These shadows and reflections are more pronounced when layers of tape are built up on top of one another to form a pattern.

3. When tape and ink are employed in the same layout, the technician is presented with a reproduction problem similar to that of reproducing a composite line original (Sec. 1.1).

Ink Method. Circuit layouts are often drawn in india or ordinary drawing ink. These present little difficulty in reproduction, since they are handled much the same as any line original drawn in ink (Sec. 1.1).

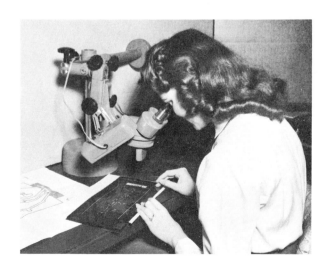

FIG. 3.2 Critical requirements demand critical measurements. A quality-control technician checks measurements of a printed circuit under a stereo-microscope. (*Courtesy of Lockheed Electronics Company.*)

Scribing Method. A glass plate, which is coated with lacquer that contains a dye for ease of visibility, is used for the scribing method. The lines of a circuit pattern are scribed into and then scraped out of the lacquer coating, leaving clear areas on the glass plate. The plate is then photographed against a white background, and a negative is made.

The scribing method produces extremely fine and accurate lines and is ideal for camera reproduction. Corrections and changes are easily made without having to rescribe the entire pattern by recoating the affected area with lacquer and repeating the scribing process in that one area only.

Peel-off Method. The material used in the peel-off method consists of two layers: a top layer of ruby-colored translucent film, such as Ruby Studnite, which is affixed to a dimensionally stable translucent bottom layer.

The layout man tapes this material over the original pencil drawing and places both on a light table. Since the ruby-colored and dimensionally stable layers are translucent, he can see the pattern of the original pencil drawing through the covering material.

With a set of sharp cutting tools, he traces the pattern onto the ruby-colored material (Fig. 3.3). Then, he peels off the background portion of the ruby-colored layer from the dimensionally stable material. Remaining affixed to the latter are the ruby-colored lines which form the circuit pattern. Any symbols or letters required on the pattern are then applied with ink or with mechanical paste-up materials.

The ruby-colored circuit pattern simplifies reproduction in the camera, since it photographs the same as ink lines and can be handled as an ink original (Sec. 1.1).

Scribe–Peel-off Method. The scribe–peel-off method of laying out an original is a photographic process. The layout man scribes the circuit pattern on the top coating of a double-coated sheet of dimensionally stable translucent material. The scribed portions of the pattern are peeled off, leaving the bottom layer of the translucent material showing through. The material is exposed to a strong light source, which results in formation of a reverse relief image of the cutout pattern on the bottom layer. A chemical

FIG. 3.3 Peel-off method. Using a cutting tool, the draftsman traces the layout circuitry on the top layer of ruby-colored translucent film, stripping away the superfluous background. The close-up shows a detail of the finished work. (Courtesy of E. I. Du Pont de Nemours and Company, Inc., and Bendix Corporation.)

process is then employed to remove the material's remaining top layer. The final product—a translucent base on which is composed a reverse relief image—is then reproduced as a line original (Sec. 1.1).

Regardless of which layout method is employed, the layout man must use a stable-base material, such as translucent polyester or polystyrene, on which to make the original. This material should have a surface suitable for inking in letters and symbols.

Stable-base materials ensure maintenance of exact scale. In contrast to regular drawing papers, stable-base materials will not change size when subjected to temperature and humidity fluctuations. Once the original drawing changes scale, it makes erroneous the final printed-circuit pattern, since all measurements for reference and final checking during reproduction are made from the original.

3.2   Camera Techniques for Printed-circuit Reproduction

The camera stage of printed-circuit reproduction can be bypassed on those occasions when an original is laid out on a transparent or translucent material, but is the size specified for the final printed-circuit board. The original is contact-printed to produce a film negative (Fig. 3.4), and the negative is then reprinted to film positive form (see Sec. 3.4). The film positive is turned over to the plating shop for production of the circuit board.

In most cases, however, the camera must be employed in reproduction of printed-circuit originals, since reduction of the original to the size required in the plating operation is necessary. The graphic-arts reproduction camera is the ideal camera for critical reduction of printed-circuit originals.

Most reproduction cameras have mechanical scaling systems for setting the correct size of a reduction (or enlargement). These scales are usually elaborate and extremely accurate micrometer settings that ensure adjustments to the minutest measurement. The reproduction camera designed specifically for holding close tolerances, and the one most reliable for very critical printed-circuit reproduction, is the metal screwdrive type.

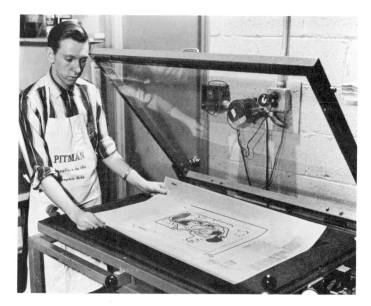

FIG. 3.4   Contact printing an original. The original has been drawn to the same size as that required for the final printed-circuit board and has been laid out on a translucent material. This work should be done in a vacuum frame to provide good contact between film and sensitive material. (*Courtesy of Lockheed Electronics Company.*)

Use of a graphic-arts reproduction camera for circuit work offers another advantage. By setting the camera for a 1:1 (same-size) copy and making a test negative, as explained below, the technician can quickly determine whether the camera in its present condition will serve his requirements. For example, if the results of a test prove that the camera is perfectly aligned, any reduction in copy will be perfect. If, however, the test negative shows that camera alignment at a 1:1 scale is off-scale a fraction, the technician can quickly determine whether this error is acceptable. *An error at a 1:1 scale is always reduced proportionately when a reduction is made.* At a reduction of 50 per cent, for example, an off-scale error of $\frac{1}{16}$ in. at 1:1 is reduced to $\frac{1}{32}$ in.; at a reduction of 25 per cent, the error is reduced to $\frac{1}{64}$ in. Specifications often permit a plus or minus allowance of this sort.

If printed-circuit work is not extremely critical, cameras other than the reproduction type may be used (see Sec. 1.2, for precautionary measures to take when reproducing with cameras other than the reproduction camera). The film size these cameras accommodate must be as large as or larger than the largest negative needed in reproduction, since an enlargement of a small negative is not practical in circuit work. Blowups lose the sharpness, detail, and accuracy usually required in such work.

No matter what type of camera is employed, the most important caution is to ensure perfect camera alignment, thus guaranteeing accurate results in exact scale reductions. (*Alignment* refers to the overall accuracy of the camera in reproducing parallel lines, so that no linear distortion is introduced.) A simple, but effective procedure you can use to check camera alignment is as follows:

1. Insert a sheet of ruled paper (the test chart) into the camera's copy easel. To ensure an accurate test, the test chart must be as large as the largest original you will be copying (Fig. 3.5).

2. Set the camera for a 1:1 copy and make a test negative using a process film normally employed in the copying of line material (Sec. 1.3). *Use only a dimensionally stable-base film* to guarantee maintenance of correct reproduction size. Materials other than dimensionally stable base could change size as temperature and humidity fluctuate, rendering the reproduction useless.

Glass plates offer the maximum in stability, but are more difficult to handle and more expensive than other dimensionally stable-base materials. Moreover, not all cameras can accommodate glass plates. Films with polyester and polystyrene bases offer less stability, but are usually stable enough for most circuit work.

3. Develop, wash, and dry the test negative according to the film manufacturer's instructions. Never force (quick) dry film used in printed-circuit reproduction, since drying by heat or with a fan causes even a stable-base material to change size. *Negatives should be air-dried at room temperature.* If a faster drying time is desired, one of two methods can be employed.

The first method is to place the washed negative on a clean, flat surface, such as a light box or metal drainboard, with its emulsion side down. Squeegee the back of the negative to remove excess water. Then hang the negative up to dry at room temperature. Squeegeeing reduces drying time.

The second method of hastening drying time is to immerse the washed

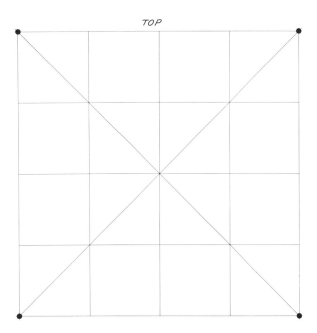

FIG. 3.5 Test chart used in checking camera alignment.

negative into a tray of alcohol for a few seconds. Then hang the negative up to air dry at room temperature. Since alcohol evaporates faster than water, drying time is shortened.

4. After the test negative is dry, place the test chart from which the negative was made on a light table *under* the negative. Check for parallel lines and exact registration vis-à-vis the test negative and test chart. If negative and chart do not match exactly and if all recommended reproduction operations were observed, it is obvious that focal plane and copy easel are not parallel, resulting in linear distortion, scale setting is incorrect, resulting in improper registration, or both conditions exist.

Linear distortion refers to the lines of the test negative not being exactly parallel to the lines of the test chart. It is caused by the focal plane of the camera and the copy easel being out of alignment. Improper registration refers to the lines of the test negative being smaller or larger in size than the lines of the test chart. It is caused by improper camera scale setting.

Correcting linear distortion in graphic-arts cameras is usually a job for the camera's manufacturer or his representative. Correcting scale can often be accomplished by the technician by refocusing his camera, thus bringing it into correct range. In general, the technician should consult the maintenance and operating guide for his camera to ascertain the recommended procedure he should follow to correct any malfunction of his equipment.

Cameras other than the reproduction type, which are adjusted by moving the copy easel and film plane, can be aligned by using a small carpenter's level to level the film plane and easel. A final alignment check can then be made by placing a piece of printed coordinate transparent or translucent paper into the copy easel. Set the camera for a 1:1 scale, and place a second piece of the same printed coordinate transparent or translucent material on the camera's ground glass. If the camera is aligned properly, the two coordinates should match exactly.

After determining camera accuracy, the reproduction technician is ready to begin printed-circuit work. One piece of equipment he will need is an accurate steel rule that is calibrated, at least, in units with which he is working. If originals for reproduction require tolerances accurate to within plus or minus ¹⁄₆₄ in., for example, the rule's scale should be calibrated at least in units of 64ths. In addition, the length of the rule should be as long as or longer than the largest negative to be measured.

A magnifier that permits the technician to see easily the minutest scale subdivision of the rule is another tool which should be kept handy.

3.3 Illumination for Printed-circuit Reproduction

Printed-circuit originals laid out with ink, or on scribing or peel-off materials, resemble a line original and are treated as such in reproduction. They can be illuminated by any of the conventional reflected light equipment discussed in Sec. 1.2.

Originals laid out with tape, however, present undesirable characteristics, as explained in Sec. 3.1, that often prohibit exposure by reflected light sources, making it necessary to employ the transillumination method of lighting.

This section describes reflected lighting and transillumination as they pertain to printed-circuit reproduction (Fig. 3.6).

Reflected Lighting. (See Sec. 1.2 for a complete discussion of reflected lighting techniques.)

Since most circuit originals are laid out on a translucent or transparent base material, a sheet of heavy white opaque paper must be placed on the copy easel to serve as backup for the original (Fig. 3.7). White paper increases the reflectiveness of the original and helps to produce a solid back-

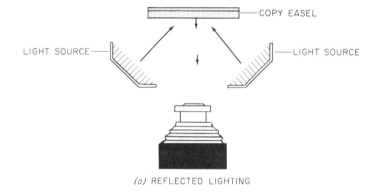

(a) REFLECTED LIGHTING

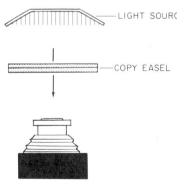

(b) TRANSILLUMINATION

FIG. 3.6 Differences between reflected lighting and transillumination.

FIG. 3.7 Reproducing a printed-circuit original by reflected lighting. Note the white background upon which the original is placed. (*Courtesy of Lockheed Electronics Company.*)

ground. In the absence of this white covering, not only is reflectiveness decreased, but any markings on the easel may show through the original's translucent or transparent base and be picked up by the camera's lens, necessitating their removal from the negative by opaquing.

Most stable-base materials designed for circuit layouts possess a set of imprinted grid lines to guide the layout man in positioning the pattern. In most cases, these are light blue in color and are eliminated in reproduction if an orthochromatic process film is employed, since the ultraviolet light rays emitted by reflected light sources make blue invisible to this type of film. If other than an orthochromatic film is used, or if grid lines are another color, the lines must be removed by the technician before reproduction, or they will record on the negative.

To remove grid lines, rub them with an alcohol-dampened cloth. If this procedure is not recommended by the manufacturer of the material, the filter method can be used instead. A filter the same color as the grid lines is placed on the camera's lens. An exposure of the original containing the grid lines is then made. The filter prevents the colored grid lines from being recorded on the negative. (For a complete discussion of the theory and applications of filters, see Sec. 5.3.)

The following is a step-by-step guide to use when reproducing circuit originals by reflected lighting:

1. Remove dust and dirt from the original, lens, and copy easel glass (if glass is used on the copy easel).

2. Center the original in the easel.

3. Set the camera for the proper reduction.

4. Make the exposure using dimensionally stable film.

5. Develop the negative in lithographic-type developer. When reproducing patterns containing intricate detail, the fine-line development technique explained in Sec. 1.1 will provide results of higher quality than usual. For

patterns that do not require special handling, conventional development techniques can be employed.

6. Rinse the negative in acetic acid, and fix in a fixing bath.

7. Wash the negative at the recommended water temperature (usually 65 to 70° F) and for the *recommended* time. Overwashing could cause the negative's base to change size.

8. Hang the negative up to dry at room temperature.

Transillumination (Backlighting). As stated before, reproduction of circuit originals laid out with tape presents problems peculiar only to this layout technique if the reproduction is made by reflected lighting. The tape's texture usually causes reflections, and the tape's thickness could cause shadows, distortion, and ragged lines on the negative. To eliminate these shortcomings, transillumination, which overcomes all reflection problems, should be employed.

Whereas reflected light is light reflected from the original, transillumination is light that passes *through* the original (Fig. 3.8). It can only be employed if the taped original is laid out on a translucent or transparent base material.

Most copy easels for a graphic-arts reproduction camera are equipped with a transparency opening. This opening is used primarily to transilluminate color transparencies when making color-separation negatives, but it can also be used when backlighting a circuit original laid out with tape. If such an arrangement is not available, a light box can be substituted.

Place the taped original in front of the camera's transparency opening (or tape it to the front of the light box). Put a sheet of frosted or opal glass between the light source and the back of the original to reduce the glare of the light (a light box should contain this type of glass in its construction). Set the camera scale, and make the exposure.

When using a graphic-arts camera, the camera's light sources for transillumination are positioned a few feet back of the transparency opening. All other lights in the vicinity of the camera must be turned off, so that all lighting that may cause reflections is eliminated. With a light box, of course, the light source is contained right in the box.

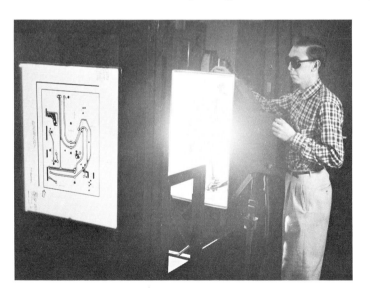

FIG. 3.8 Reproducing a printed-circuit original by transillumination. The transparency opening of the graphic-arts reproduction camera's copy easel is used. Note the fine detail this type of lighting brings out. (*Courtesy of Lockheed Electronics Company.*)

Transillumination exposures are usually 2 to 2½ times longer than normal reflected light exposures. It is advisable to perform tests to establish correct exposure time.

The transillumination technique could present a problem if a portion of the printed-circuit original, such as lettering and symbols, is drawn in ink, while the remainder of the pattern is laid out with tape. All ink areas should be checked by the cameraman for opacity, that is, the ink's ability to hold back light. Check the original circuit pattern on a light table to determine whether light, which could be recorded on the film, is passing through ink characters. All weak-inked areas should be touched up before reproduction with opaquing fluid or india ink, and the layout man should be advised to use a more opaque ink when putting ink lettering and symbols on a taped original.

The following is a step-by-step guide to use when reproducing printed-circuit originals by transillumination (additional information on trans-illumination is presented in Sec. 5.4):

1. Remove grid lines on the material on which the layout is made, as previously explained.

2. Check the circuit original over a light table to ensure opacity of all inked areas.

3. Line up the original in the transparency opening of a graphic-arts camera or on a light box.

4. Position the illumination source in back of the transparency opening, if a graphic-arts camera is employed. The light box supplies its own light source.

5. Set the camera for the proper reduction.

6. Make the exposure, using dimensionally stable film.

7. Develop in a lithographic developer.

8. Rinse the negative in acetic acid, and fix in a fixing bath.

9. Wash the negative at the recommended water temperature (usually 65 to 70°F) and for the recommended time.

10. Hang the negative up to air dry at room temperature.

If the equipment needed for transilluminating taped originals is not available, use a dimensionally stable autopositive material to make an intermediate. An autopositive image—that is, a positive image produced without going through the negative stage—is made, permitting you to avoid reflections encountered with reflected lighting by converting the tape layout to a black-silver image that will not produce reflections and glare when subjected to reflected light sources.

This is done by contact printing the taped original onto the autopositive material. Thus, overlapping tape areas of the original are reproduced as black solid areas on a white background, although the overlaps and raised surfaces are plainly visible on the original. By eliminating lighting setups, where two lights crosslight the original, causing shadow areas around the overlaps, you eliminate the problem of reflection.

The autopositive intermediate, which can be handled in reproduction as a line original drawn in ink (Sec. 1.1), is then photographed by reflected light to produce the final negative. For a complete discussion of the auto-positive technique, see Chap. 8.

FIG. 3.9 Ensuring accuracy. The reproduction and plating technicians check a film positive against a negative before the positive is sent to the plating shop for conversion to a printed-circuit board.

3.4 The Film Positive in Printed Circuitry

It is customary for a film positive to be used by the plating shop in the final production of printed-circuit boards. Making the film positive is the responsibility of the reproduction technician, who should check with the plating shop to determine exact requirements (Fig. 3.9).

The film positive is made directly from the negative by simply contact printing the negative image onto a sheet of lithographic film, which is developed in a lithographic developer. Certain innovations can be employed, however, that add to the quality of the positive.

The highest-quality film positive—one that offers extreme accuracy and sharpness—is obtained by using a vacuum frame and a pinpoint light source. Most vacuum frames build up between 15 and 30 lb of pressure, which is enough to ensure perfect contact between the negative and the material on which the film positive is being made. Absolute accuracy of the final positive can be obtained only through perfect contact.

To make a film positive, place the negative in contact with the sheet of unexposed lithographic film, so the unexposed film is on the bottom and the negative faces the light source. The negative and film are sandwiched together, emulsion to emulsion.

Insert the negative-film sandwich in a vacuum frame having a glass cover. When the vacuum builds up to the desired point, make the exposure, preferably with a pinpoint light source. The exposed sheet of lithographic film is then processed according to the manufacturer's instructions.

A pinpoint light source can be purchased from a graphic-arts equipment dealer, or it can easily be constructed by converting a safelight housing. The ideal place to mount the source is in the ceiling, directly over the vacuum frame (Fig. 3.10).

To make a light source, use an Eastman Kodak 5-in. safelight lamp. Insert a black cardboard or metal mask into the safelight's filter slot, and cut a small hole in the center of the mask so that the light is emitted as a narrow beam. A hole ¾ in. in diameter provides a pinpoint light beam sufficient to cover a 30- by 40-in. area from a ceiling distance of about 6 to 8 ft.

As the light source itself, use a Westinghouse projection bulb No. 100 T8/1 SC. This bulb is rated for use at exactly 20 volts, making the employment of a voltage regulator or a Variac a requirement. If the bulb is burned at higher than 20 volts, it will shatter.

A timer is used in conjunction with the voltage regulator or Variac to control exposure time. The light source is plugged into the voltage regulator, and the regulator, in turn, is plugged into the timer.

A pinpoint light source is an excellent tool to use for making fine-detail contact exposures. The pinpoint source is better suited for fine-detail reproduction than wide-source lights, since the results are sharper with less undercutting of the image.

3.5 Selective Photoetching

Lettering or identification symbols contained in a very intricate circuit pattern may present an electrical problem. If letters or symbols are placed between two conductors on a circuit with little separation between them, an arc could form across the symbols or letters, causing a short in the circuit.

To avoid this possibility, you can assist the plating technician in the use of selective photoetching, which is the employment of a silk screen to form letters and symbols in ink or paint on the final circuit board. Thus, letters and symbols would not be in the form of metal, eliminating entirely the chances of arcing.

If silk-screen stencils are to be used (see Chap. 4), the reproduction

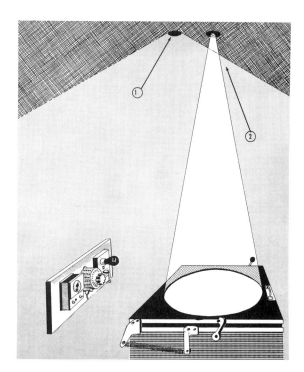

FIG. 3.10 A typical contact printing setup. The arrangement can be used for other types of contact work besides printed circuitry by mounting a more intense light source in the ceiling, as shown by (1) on the diagram. The pinpoint light source is shown by (2) on the diagram. The two lights operate off separate switches; so they do not conflict with each other.

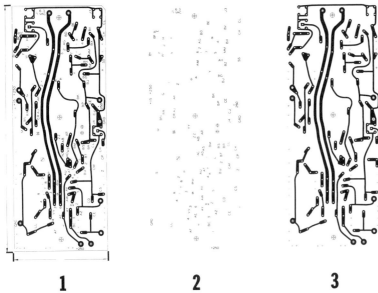

1 **2** **3**

FIG. 3.11 A sample of selective photoetching. Illustration 1 shows the original circuit layout, including pattern, symbols, and lettering. Illustration 2 typifies the silk screen. This is placed over the final metal printed-circuit board, with the letters and symbols applied to the board with ink or paint by brushing over the screen. Illustration 3 shows the circuit pattern with letters and symbols removed.

technician must make two negatives of the original layout (Fig. 3.11). He opaques out all letters and symbols from one of the negatives, leaving only the conductor pattern and registration marks. He then opaques out the circuit pattern from the second negative, leaving only the symbols, letters, and registration marks.

The two negatives are then contact-printed separately to film positive form. The positive displaying only the circuit pattern is used for the production of the final board (in metal). The second film positive displaying symbols and lettering is transposed into a silk-screen stencil, which is placed over the final metal board. Symbols and letters are then applied in ink or paint simply by brushing over the stencil. (A complete discussion of silk-screen techniques is presented in Chap. 4.)

4 | SILK-SCREEN STENCIL PREPARATION

SCREEN PRINTING is a versatile process that enables one to print designs and images, including line and halftone images, on *any type of surface,* regardless of the size, shape, or material composition of the surface.

The silk-screen process permits printing of subject matter that cannot be printed by conventional press processes. For instance, one cannot print words, drawings, line images, or halftones on milk bottles, toothpaste tubes, or rubber balls with a conventional press; but this work is child's play for a silk-screen stencil. Most presses cannot accommodate large-size billboard signs. Yet, silk-screen stencils can be made to any desired size. Furthermore, printing printed-circuit patterns and metal nameplates are jobs that can be done efficiently only by screen stencils.

A silk screen is nothing more than a stencil that has open and closed areas. Although it is implied that the stencil is made of silk, many other materials, such as wire, nylon, and Dacron, are often used instead.

In most cases, the screen image is a porous, open area, while the screen background is solid and nonporous. In applying the image to a surface, ink or paint is forced through the porous opening by running a rubber squeegee back and forth over the stencil. The ink or paint cannot penetrate any area but the image, since all other areas are nonporous.

In other cases, as explained in Sec. 3.1, a reverse image may be desired. Consequently, the screen image is prepared as the nonporous, solid area, preventing ink or paint from penetrating on application, while the background area is made porous and open.

Some may contend that the subject of screen stencils belongs in a book written for the printing industry, and not the reproduction industry. Preparing and using screen stencils for transferring an image onto a surface is, after all, a printing process.

In recent years, however, silk-screen printing has assumed new importance to those engaged in reproduction. The main reason for this condition, perhaps, has been the perfection of photographic materials for stencil preparation. Unlike those in printing who specialize in large-scale production of screen stencils, the reproduction technician is concerned primarily with using these materials to prepare stencils which meet the individual requirements of his customers or, if he is employed in industry, his company. By associating silk screening in this limited sense with reproduction, the services the technician can now offer customer or company are greatly broadened, and new areas of business are opened.

Large-scale production printing by silk screen, such as that employed in printing images on rubber balls, toothpaste tubes, billboard signs, and the like, are handled by special machines capable of high-volume output. Images reproduced on silk screen by a reproduction technician, on the other hand, are applied to a surface by hand, since the screen is usually employed in printing a limited number of items.

There are three screen preparation methods considered within the realm of the reproduction technician. One, the so-called carbon tissue method, has been adopted by reproduction laboratories from the screen printing industry. The other two, the Eastman Kodak Ektagraph method and the Du Pont screen process film method, are processes inherent primarily to the reproduction industry.

Although each of the methods differs in practice, the basic principle of all three is the same; that is, an image is exposed on a sensitive material and is developed. In this respect, *the work is reproduction.* The developed material is then positioned on and adhered to a screen.

Carbon tissue, Ektagraph, and screen process film permit the reproduction technician to fit screen-stencil preparation into his setup without an increase in facilities or equipment. The remainder of this chapter discusses in detail preparation of screen stencils by all three methods.

4.1 The Carbon-tissue Method

The carbon-tissue method (also called the pigmented paper method or the dichromated gelatin method) requires an ultraviolet light source of very high intensity to expose the sensitized material. Exposures are usually of long duration, making this a slow process as compared to others. For this reason, the carbon-tissue method is not recommended for preparing stencils that will be used in printing fine detail, since long exposures tend to block up the reproduced image. A major advantage of the carbon-tissue method is that all steps may be carried out under room illumination, with no darkroom procedures necessary.

The production of stencils with pigmented paper has been employed by the screen-process printing industry for over 100 years. Yet, it is still widely

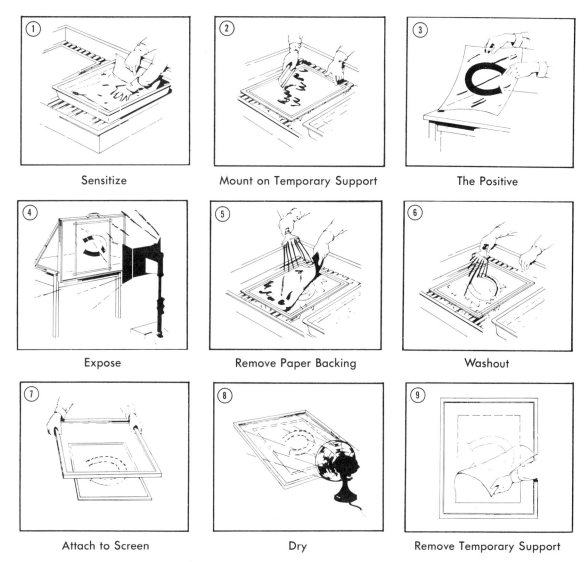

Sensitize	Mount on Temporary Support	The Positive
Expose	Remove Paper Backing	Washout
Attach to Screen	Dry	Remove Temporary Support

FIG. 4.1   The carbon-tissue method. (*Courtesy of McGraw Colorgraph Company.*)

used. Because of its simplicity and economy and the excellent results it produces, the method has been adopted by the reproduction industry and has fit into the reproduction setup without any increase in equipment.

The term *carbon tissue* refers to the carbon black that used to be added to the gelatin in the pigmented paper to make a visible image. The term is now a misnomer, since present-day materials do not require a carbon additive. Instead, the paper is coated with gelatin and a pigment. When developed, the color of the pigment forms the visible image. This is the reason why the process is also referred to as the *pigmented-paper method.*

There are nine steps in preparing a stencil with the carbon-tissue method, as follows (Fig 4.1).

Step 1—Preparing the Positive Original. The composition of the final screen image depends on the composition of the film positive image to be reproduced on the pigmented paper. If the positive image is a line original composed of solid, opaque areas and a clear (or "white," as most technicians call it) transparent background, the ultraviolet light used in exposure cannot penetrate the solid area, but can penetrate the clear area. Areas of the gelatin-coated paper not penetrated by light remain unhardened and will wash away. Areas affected by light are made hard and will not wash away.

On the final stencil, then, the washed-away gelatin area (the area covered by the solid positive image) becomes porous and open, permitting ink or paint to pass through. The gelatin-hardened area (the area covered by the original's clear background) becomes nonporous and solid, blocking the passage of paint or ink.

The opposite is true if a reverse image is to be placed on the final stencil. In this case, the image area of the original positive is clear and the background is black. Ultraviolet light penetrates the clear area during exposure, producing a nonporous solid area on the sensitive material, while no light penetrates the solid background. When printing with this type of stencil, paint or ink is forced through the background, but not through the image area, thus producing an outlined image.

The film positive image may be fashioned in several ways for exposure to the pigmented paper. It can be prepared, for example, with opaque black drawing ink, opaque paste-up letters, ruby-red plastic-masking materials, or amber-colored masking acetate. Little difficulty is experienced in reproducing line originals of this nature on pigmented paper.

The film positive could also be a halftone, which could present difficulty in reproduction. Caution must be exercised in reproducing the fine detail of a halftone's dot structure, although coarser screens, such as 65-line, offer minor problems. In general, the finer the halftone positive screen, the more critical the work becomes.

Before positive image (line or halftone) and sensitized pigmented paper are brought into contact, carefully inspect the positive for pinholes in the image area. Touch them up with opaque ink or opaquing fluid. Dust specks on a positive's clear transparent areas may also cause trouble on the final stencil and should be removed by brushing the positive with a draftsman's brush or a soft chamois cloth.

Step 2—Sensitizing the Pigmented Paper. Unsensitized pigmented paper is the only type available from graphic-arts dealers. The paper's emulsion (gelatin) must be made light-sensitive by the technician before reproduction. This is accomplished by immersing the paper into a solution of potassium dichromate. (You can also use ammonium dichromate or sodium dichromate, but these are more expensive and are no more effective.)

Distilled water is recommended for mixing chemicals, although tap water may be used in its absence. All containers for mixing and storing the dichromate solution should be absolutely clean and used only for this chemical.

A 2 per cent solution of dichromate is recommended for the best and safest results. A more concentrated solution produces a higher sensitivity, increasing the paper's speed, but is more difficult to work with. Dichromate causes skin irritations. Even with 2 per cent solutions, rubber gloves should be worn when preparing and using the chemical.

Solution temperature should be between 45 and 68°F, with a temperature of 55 to 57°F as the optimum. It is often difficult to maintain this temperature range, especially in hot weather. You can overcome effects of heat, however, by replacing about 20 per cent of the water used in mixing the solution with an equal amount of isopropyl alcohol. The lower solution temperature afforded by alcohol reduces swelling of the paper's gelatin and eliminates the possibility of the gelatin melting during exposure.

Dichromate solution temperatures can also be kept within the recommended range by using water from a temperature-controlled sink to mix the chemical or by storing prepared chemicals in a refrigerator until needed. The life of the dichromate solution is prolonged by storage in a refrigerator.

Swelling of the paper's gelatin coating affects emulsion sensitivity and speed. High temperatures make the emulsion overswell, resulting in loss of speed. This excess swelling may also affect adversely adherence of the image (reproduced on the pigmented paper) to the screen. Low temperatures, however, tend to reduce or prevent swelling, thereby increasing the speed of the emulsion and making reproduction of image detail possible.

The length of time the pigmented paper is kept in the dichromate sensitizer also has a bearing on swelling and, consequently, on sensitivity. Careful control of this period is important. Immersion times vary, usually from 1 to 4 min, with the type of paper. Therefore, the manufacturer's recommendations should be followed.

Fill a tray with about ½ in. of dichromate for immersing small-size sheets of paper (up to 11 by 14 in.), and with about 1 in. of solution for large-size sheets (larger than 11 by 14 in.). Put the pigmented paper into the solution with its emulsion side up, and rub the emulsion with an elastic sponge or the palm of the hand to eliminate air bells. The paper's emulsion must be completely submerged at all times during the sensitizing process.

You can usually retain dichromate solution for about 10 days or when 60 sq ft of paper per gallon of solution has been sensitized, whichever comes first. If alcohol is added to the solution in hot weather, discard dichromate immediately after use. Alcohol evaporating from the solution changes the chemical balance of the dichromate.

Once pigmented paper is sensitized, it should be handled under subdued light. Paper can be safely exposed to low-level tungsten illumination (60- to 100-watt bulbs) for up to 2 hr. It can be handled safely for any period under bright yellow light, since the sensitized emulsion is not responsive to yellow. However, caution must be taken not to leave the sensitized paper under the ultraviolet rays of fluorescent illumination for too long a period, since the emulsion will become exposed.

Step 3—Attaching the Temporary Support. While the sensitized paper is still wet from the dichromate solution, a temporary, sturdy support is attached to make the paper easier to work with. The temporary support can either be a clean, clear sheet of plastic (or heavy acetate) or the original film positive itself.

If plastic or acetate is used as the support, place it on a flat, cool surface, and sponge it down with dichromate solution before adherence of the pigmented paper. If the film positive is used, dip it into a vinyl solution, which is available from graphic-arts dealers. The vinyl-coating protects the emulsion and image of the positive during adhesion.

After wetting down plastic or film positive, sandwich it together with the dichromate-saturated paper. If plastic or acetate is used as the support, place the pigmented paper on the plastic *with the paper's emulsion side down.* If the film positive is used, place the emulsion of the pigmented paper *in contact* with the emulsion of the film positive.

Once the paper is in contact with plastic or positive, lightly stroke the back of the paper with a squeegee. This removes air bells and flattens the paper. After the paper is perfectly flat, use even, heavy strokes of the squeegee to adhere the paper firmly to the temporary support. Now, wipe off excess sensitizer from the back of the pigmented paper, turn the sandwich over, and dry the back of the support by blotting with paper towels or newsprint.

Du Pont's Mylar film is an excellent material to use as a plastic support. That film which is 0.002 in. thick is recommended for general use. If stencils are to be larger than 30 by 40 in., however, use the Du Pont Mylar film that is 0.005 in. thick. For stencils that are to contain relatively fine detail, Du Pont's Mylar film (0.001 in. thick) assists in minimizing the tendency to undercut and block up fine lines.

Step 4—Exposure. Using the film positive as a temporary support offers shops not possessing a vacuum printing frame a distinct advantage. With a plastic or acetate support, a vacuum frame is needed during exposure to bring pigmented paper and film positive image into the closest possible contact. But when using the film positive as a support, it is already in the closest possible contact with the pigmented paper, making the use of a vacuum frame unnecessary. Exposure is made by taping a paper-film positive sandwich to a flat surface in front of the ultraviolet light source with opaque masking tape. The positive image (now the temporary support) is nearest the light source.

Use black masking tape to mask a narrow border on all four sides of the sandwich. This provides a "safe" area of unexposed emulsion on the pigmented paper, which is necessary for proper separation of the reproduced image and paper backing during washing (see step 5 below).

When a sheet of plastic or acetate is used as the temporary support, use a vacuum frame for making exposures; there is then an adequate contact between the positive image and the pigmented paper-plastic backing sandwich. Place a sheet of cardboard on the rubber blanket of the printing frame, and put the back of the pigmented paper, still wet from dichromate, against it. The cardboard prevents the blanket from being damaged by the dichromate.

With the plastic-paper sandwich in place, lay the film positive image in contact *with the plastic backing,* with the emulsion side of the image in contact with the plastic. Mask a border on all four sides of the setup, as explained above.

Close the vacuum frame, and build up suction to 15 lb per sq in. or 15 in., depending upon the way your frame is calibrated. Be extremely careful in applying pressure. Excessive vacuum can squeeze the pigmented paper's soft gelatin, which results in a poor stencil. Make a few preliminary tests to determine ideal vacuum pressure.

As mentioned before, sensitized pigmented paper is sensitive only to high-intensity ultraviolet light. Many types of ultraviolet sources are suitable for exposure, including carbon-arc lamps, mercury-arc lamps, banks of fluorescent lights, and sun lamps. Of these, the carbon-arc lamp is probably the most reliable and the most widely used, since it requires less time to expose the image than other sources.

No matter what the light source, be certain the light is centered in front of the setup to ensure even lighting over the entire area to be exposed.

Heat is a constant problem during exposure, since many types of light sources create an excessive amount. Offset the effects of heat by positioning an electric fan to blow across the printing frame or across the surface to which the setup is taped.

Different light sources require different exposure times, depending on the intensity of the source, the distance of the light from the setup, and the sensitivity of the pigmented paper's emulsion. You should make exposure tests, following the paper manufacturer's instructions. Once you have determined proper exposure for one stencil, however, work from then on with the same materials and lights is simplified.

The following are recommendations concerning each of the more popular ultraviolet light sources.

Carbon-arc Lamps. Use a *single* carbon-arc lamp for making screen stencils. The double set of carbon-arc lamps most often used in reproduction laboratories is not suitable, since the two light beams hitting the sensitized emulsion at different angles could cause a double (ghost) image because of the thickness of the stencil.

Fluorescents. Arrange banks of fluorescent lights about 4 to 6 in. apart and 7 to 12 in. away from the sensitive material. Use blue-white or blue bulbs, as these are high in ultraviolet.

Sun Lamps. Standard sun lamps of the screw-in type when placed about 2 ft away from the setup provide good coverage of an 11- by 14-in. area. They also afford reasonable exposure times (up to 10 min).

Mercury-arc Lamps. Mercury arcs usually require 2 to 3 min to warm up. Cover the easel until the lights reach maximum intensity, thereby preventing exposure of the pigmented paper until ready.

Step 5—Removing the Paper Backing. After exposure, the reproduced image must be "developed." In reality, this image is the stencil and must be separated from the pigmented paper backing.

Put the exposed pigmented paper, still attached to its temporary support, on a piece of heavy glass or Plexiglas, which is placed so that water flowed over the glass drains into a sink. Place the exposed material on the glass with its emulsion (temporary support side) facing down, and with the backing of the paper facing up.

Flood the entire backing with hot water at 115 to 125°F. Use a rubber hose attached to a shower-type sprinkler. As soon as the paper's emulsion starts to bleed (run) along the edges, carefully peel off and discard the paper's backing. What remains is the stencil image still attached to the temporary support.

Step 6—Washout. Continue to flood the stencil with hot water until all clear, open areas are clean and free of gelatin. Then, lower the water temperature to about 70°F, and continue to wash for another 2 min.

The hot water removes all soluble gelatin which was unaffected (unhardened) by the ultraviolet light during exposure. These are the areas that were covered by the solid, opaque parts of the positive original. After washout, these areas become the porous parts of the stencil. Usually a wash time of from 5 to 10 min is sufficient to wash away all unhardened gelatin.

Step 7—Attaching the Stencil to the Screen. Place the wet stencil, emulsion up, on a piece of flat material, such as ¼-in. tempered masonite. The masonite should be slightly larger than the stencil image, but smaller than the overall frame to which is attached the screen. (For a discussion of how to make a screen frame, see Sec. 4.4.)

Hold the screen, frame side up, over the stencil, and align the two. Carefully lower the screen into contact with the wet stencil. Now, remove excess moisture from the screen by blotting with a lintless blotter.

Step 8—Drying. Never dry a screen stencil made by the pigmented-paper method with heat, since heat prevents the stencil's gelatin from properly adhering to the screen. Dry with an electric fan blowing across or directly on the screen.

Step 9—Removing the Temporary Support. After the stencil has dried completely, remove the temporary support by peeling it off. Start at one corner and gently pull diagonally across the stencil. The support should come off easily. If not, the stencil is not completely dried or the support material was not perfectly clean before it was adhered to the pigmented paper.

After the support has been removed, any part of the stencil having unwanted open areas, such as pinholes, should be blocked out by using commercially prepared blockout solution.

4.2 The Eastman Kodak Ektagraph Film Method

Ektagraph is a photographic film used in making stencils for screen printing. It is a versatile material that produces excellent quality stencils in a minimum of time.

You do not need a darkroom facility when working with Ektagraph, but the film must be handled under subdued light that is low in ultraviolet. A tungsten bulb no larger than 60 watts, placed not less than 5 ft from the film, is recommended. Never expose the film directly to light before exposure and processing, and never expose it to fluorescent light or daylight under any circumstances, since these are high in ultraviolet.

Ektagraph film is an excellent material for reproducing fine detail. It is used for contact exposures only.

Essentially, the basic principle involved in screen preparation using Ektagraph is the same as with the carbon-tissue and Du Pont screen process film methods. After Ektagraph is exposed and processed, the film is transferred to a screen and is used as a stencil.

Begin with a good photographic line or halftone film positive that has been produced by conventional lithographic methods or a contact auto-positive on thin base material.

In the case of halftones, the positive should be made in a manner similar to those produced for the lithographic process, but should have pinpoint dots in the highlights, and the shadows should be barely closed in. A half-tone screen with approximately 65 lines per in. should be used when pre-paring halftones for transfer to Ektagraph. Coarse screens are recom-mended, since halftones reproduced on a silk screen require unusual care in selection of inks, printing surfaces, and techniques.

You can also expose Ektagraph film directly from artwork drawn on translucent material, such as tracing paper, cloth, and linen, provided inked areas are thoroughly opaque.

The major steps involved in producing screen stencils with Ektagraph film, and the approximate times required for each step, are: exposure, 40 sec; processing, 105 sec; washoff, 60 sec; transfer to silk and drying, 5 min; stripping off the base, 3 min. Each of these steps is explained below.

Step 1—Exposure. Kodak Ektagraph film is exposed through the base of the film, which is the side (nonemulsion) having the lighter shade. Place the emulsion side of the positive original against the base side of the Ekta-graph film.

Now, put the sandwiched original and Ektagraph in a vacuum or pressure-back printing frame with the positive image toward the glass of the frame, that is, with the original nearest the light source. Ektagraph is exposed through its base side.

Using a No. 2 floodlamp 4 ft from the setup, expose for from 30 to 50 sec if exposing a clear film positive (Fig. 4.2). When transferring images made on other materials, such as an original drawn on linen or an auto-positive, make a test strip to determine correct exposure time. Combina-tions of line and halftone copy on the same original may be exposed at the same time.

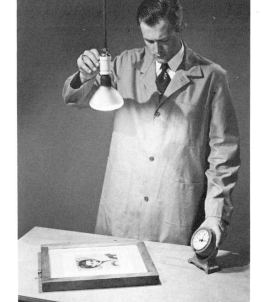

FIG. 4.2 Ektagraph method: exposure. (*Courtesy of Eastman Kodak Company.*)

Step 2—Processing. Ektagraph film contains the correct type and amount of developing agent right in its emulsion. This agent develops the image when the film is immersed in Kodak Ektagraph activator solution.

Use the solution in a tray, without dilution, at a temperature of between 65 to 75°F. If the film has been exposed correctly, adequate development takes from 1 to 1¼ min. Never remove the film from the tray while developing. Agitate by rocking the tray back and forth. The image as seen through the film's base side becomes completely black about 15 sec after development is terminated.

Immediately after development, place the film in a tray of Kodak Ektagraph stop bath A at 65 to 75°F, and agitate for 30 sec. Then immediately transfer the film into Kodak Ektagraph stop bath B at 65 to 75°F, and agitate for another 30 sec. (All chemicals mentioned thus far are available in prepared form from your dealer.)

You can use Ektagraph activator and stop baths to process about 20 sheets of 8- by 10-in. film. Discard all solutions at the end of each day, since chemical changes occur that affect their composition.

Step 3—Washoff. Remove the film from the second stop bath and place it, emulsion up, in a sink on a sheet of glass, Plexiglas, or Lucite. Play a gentle stream of warm water (90 to 100°F) over the film, using a spray head attached to a hose. The water should not exceed 100°F in temperature, or you may encounter adhesion problems when applying the film to silk. Unexposed portions of the film are usually washed off in less than a minute, leaving the area completely clean (Fig. 4.3).

The stencil is properly prepared when all the film's undeveloped cream-colored emulsion has been removed by the water spray. With line images, the detail should look clean and open. Underexposure results in tanning of the image, which results in a loss of some of the fine detail during washoff. Overexposure or an insufficient wash results in unclear and slight closing up of the reproduced fine lines.

FIG. 4.3 Ektagraph method: washoff. Note the clear appearance of the halftone reproduction. (*Courtesy of Eastman Kodak Company.*)

The dots of a reproduced halftone image should be sharp and clear. Overexposure or an incomplete wash presents a veiled or mottled muddy appearance in the dark areas when viewed over the transmitted light of a light table. Underexposure results in the loss of the smallest gelatin dots in the washoff step.

After washoff, the film is ready for transfer to silk. *Do this immediately.*

Step 4—Transfer to Silk and Drying. The silk used for the finished stencil must be absolutely clean before the image is applied. If using a new piece of screen material, mount it on its frame (as explained in Sec. 4.4), and clean it to remove oil and sizing material acquired during manufacture.

Wash a new screen with cleaning solvent, followed by a wash in detergent and a diluted solution of sodium hypochlorite (1 part of laundry bleach to 9 parts of water).

If re-using an old screen, ink and gelatin of previous stencils must be removed so that the new stencil adheres properly. Soak the screen in hot water (about 125°F) for several minutes. Then swab with Kodak etch bath EB-1 until the stencil is softened. Make the etch bath solution by using the following formula:

Part A

Material	Avoirdupois or U.S. liquid measure	Metric measure
Water.....................	12 oz (liquid)	375 cc
Cupric nitrate..............	6¾ oz (powder)	200 grams
Kodak potassium bromide.....	145 grains (powder)	10 grams
Kodak glacial acetic acid*....	5 oz (liquid)	160 cc
Water to make.............	32 oz	1 liter

* With Kodak acetic acid (28 per cent), use 18 oz (570 cc).

Part B. 3 per cent hydrogen peroxide (fresh, commercial grade). For use, mix equal parts of A and B.

Following etch bath application, scrub the screen surface lightly with a scrubbing brush while rinsing with hot water. Finally, sponge the screen with a diluted solution of sodium hypochlorite (1 part of laundry bleach to 9 parts of water), and wash with water.

To transfer stencil to clean silk, place the wet film, face up, on a platform (a clean porcelain tray turned upside down will suffice). This platform should be smaller than the wooden frame containing the silk, since the frame is to be placed over the platform. It should, however, be larger than the reproduced image area.

Make sure the silk is wet. With the bottom of the frame facing up, carefully place the silk down on the Ektagraph. Weight the frame with some light objects to hold it in place, but be careful not to exert any pressure on the silk itself.

Blot up excess water on the screen with newsprint or paper towels and begin the drying operation (Fig. 4.4). Drying can be quickened by using a fan or by combining a fan with a heat-producing infrared lamp. If using a heating lamp, place it a few feet from the screen, and put the fan so that it

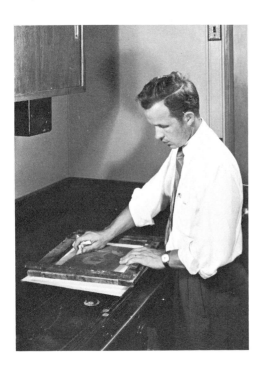

FIG. 4.4 Ektagraph method: blotting in preparation for drying. (*Courtesy of Eastman Kodak Company.*)

blows across the top of the silk. When the exposed surface of the silk is thoroughly dry, turn the frame over and dry the stencil's back.

Step 5—Stripping Off the Base. Once the stencil is dry, remove the Ektagraph base. Liberally apply Kodak Ektagraph stripping solvent to the silk side of the stencil. Turn the frame over, and note the progress of the stripping action by observing it through the base of the film. Fine-detail areas turn from black to gray during stripping.

After 3 min of keeping the silk side moist with stripping solution, the film base can be pulled away (Fig. 4.5). After the base is stripped off, allow solvent to evaporate completely from the stencil. Finally, mask the edges of the silk screen. It is now ready for printing.

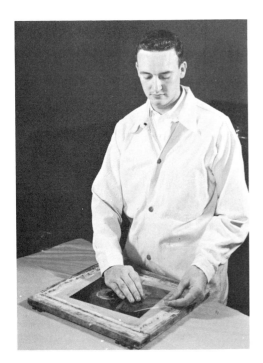

FIG. 4.5 Ektagraph method: stripping off the base. (*Courtesy of Eastman Kodak Company.*)

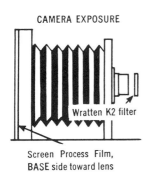

CAMERA EXPOSURE

Wratten K2 filter

Screen Process Film, BASE side toward lens

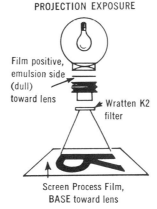

PROJECTION EXPOSURE

Film positive, emulsion side (dull) toward lens

Wratten K2 filter

Screen Process Film, BASE toward lens

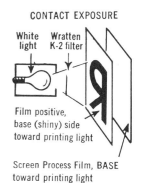

CONTACT EXPOSURE

White light   Wratten K-2 filter

Film positive, base (shiny) side toward printing light

Screen Process Film, BASE toward printing light

FIG. 4.6 Exposure of Du Pont's screen process film. (*Courtesy of E. I. Du Pont de Nemours and Company, Inc.*)

4.3 The Du Pont Screen Process Film Method

The Du Pont screen process film method offers wide latitude to the reproduction technician engaged in making screen stencils. The material can be exposed in three ways: by contact, by projection, or in a camera (Fig. 4.6). Darkroom procedures are required in processing the film.

Screen process film is an ideal material to use for making screen stencils from original artwork, from undersize negatives or positives that must be enlarged, or from oversize negatives or positives that must be reduced. This film is especially suited for the small shop that does not have a large copy camera.

Du Pont screen process film is a high-contrast, silver-sensitized orthochromatic photographic material on a dimensionally stable base. Its high speed, as compared to other photographic stencil material, permits exposure by projection with conventional photographic enlarging equipment. After exposure, processing, and laying the stencil on a screen, the stable film support is stripped off and the emulsion layer becomes the actual stencil for screen printing.

Generally, there are three steps involved in making screen stencils with screen process film: exposure, darkroom processing, and white-light processing.

Step 1—Exposure. During exposure *and processing,* the film may be handled safely under a red safelight, such as a Wratten 1A, at a distance of no closer than 4 ft. All other types of light fog the material.

No matter what exposure technique is used—camera, projection, or contact—the final result is the same. A properly exposed stencil develops to a dark brown color (not black) when viewed over a light box. Generally, the maximum exposure recommended is one just before the finest lines begin to fill in. A densitometer reading of a properly exposed background area should read between 1.70 and 1.95.

Exposure by Camera. You can expose screen process film in a camera directly from the original copy. The film must be mounted in the camera with its *base toward the lens.*

You may use a pressure back on the camera, provided the film is held flat and firm during exposure. However, a camera with a vacuum back is recommended. If a camera with a "sticky" back is used, the back must be covered with a clean film base or cellophane to prevent the film's emulsion from contacting the "sticky" surface. The film can then be mounted with pressure-sensitive tape.

Since screen process film is exposed by yellow light, a Wratten K2 or equivalent yellow filter must be used over the camera lens. This filter permits yellow light to pass through to the film, blocking out other light, especially ultraviolet, which fogs the material. Until you gain experience in making camera exposures with screen process film, make test strips at various times or lens openings to arrive at correct exposure.

The need for accurate exposure is extremely critical when an original is in poor condition or contains fine lines. To improve contrast and to ensure that fine lines are not lost when transferring the original image to screen stencil, copy a poor original on an orthochromatic lithographic film (see Chap. 1), and make a positive intermediate. Then copy this intermediate, which is of higher contrast than the original, onto the screen process film.

When setting up the original for camera copying with screen process film, mask off all extraneous white areas around the original with black paper, thus eliminating reflections. Keep lens and filter free of dust and fingermarks, and use a lens shade over the camera to keep out stray light.

Exposure by Projection. You can project a positive transparency or slide onto screen process film with an enlarger, preferably of the condenser type. Use a yellow filter (Wratten K2 or equivalent) to provide a yellow-light source for exposure.

Put the transparency or slide into the enlarger carrier with its emulsion side down. Place a sheet of screen process film into the enlarger easel with its *base up toward the lens.* After development, the result is a reverse-reading image on the film's base.

Determine correct exposure by making test strips. A typical exposure for an eight times magnification with a condenser-type enlarger using a No. 212 photoenlarger lamp is 60 sec at $f/8$.

Take precautions to guard against light leaks and reflections while projecting. One method is to use a piece of black paper or other nonreflecting material on the enlarging easel to prevent halation and to maintain image sharpness.

Most enlargers and projectors leak some white light around the lamphouse and negative or slide carrier. Protect film from this light by placing black paper over areas that leak light.

Exposure by Contact. If possible, use a vacuum printing frame. You can use a pressure frame if a good, even contact between copy and film is maintained.

Place the screen process film in the printing frame with its *base toward the light source.* Place the original positive's emulsion side in contact with the film's base side.

Yellow light must be used for exposures. A white, frosted 25-watt lamp enclosed in a lighttight lamphouse with a 1½-in. opening that is covered with a yellow filter (Wratten K2 or equivalent) is ideal. A typical exposure with the light held 5 ft from the printing frame is 25 sec. If bulbs of higher wattage are used, the distance from bulb to printing frame must be increased or exposures decreased. Wattage, diameter of the lamphouse, and distance from light to film should be controlled to give an exposure time of not less than 20 sec.

Enclosure of the light source for contact printing is recommended since the smaller opening gives a much sharper printing light and provides maximum exposure latitude. The bulb should be placed as close to the opening of the lamphouse as possible to ensure an even light spread over the entire printing frame. Use an adapter, if necessary, to extend the socket.

Step 2—Darkroom Processing. Du Pont screen process film is developed in two commercially prepared chemicals: Du Pont developer A and activator B. Keep solution temperatures within a range of 65 to 75°F. All processing must be done under a red safelight, such as a Wratten IA.

Place the exposed piece of film in a tray of developer A. Wet it thoroughly to remove air bells, and agitate frequently for 1 min. No image appears on the film while it is in the solution.

Remove the film from developer A and drain for at least 15 sec. Then, put it into a tray of activator B, with emulsion *down,* and agitate to prevent air bells. Turn the film over and agitate moderately for 1 min. This procedure blends the carry-over of developer A with activator B and helps to form the image. Be certain that the film is fully covered with solution at all times during this step.

Now, remove the film from activator B, *without draining* and rinse immediately in a tray of running water for 20 to 30 sec. Transfer the film quickly between the B solution and the water rinse, and again between the water rinse and the acid stop bath that follows. Excessive exposure to air at these stages can cause aerial oxidation, which interferes with proper washout of the stencil.

Rinse the film in a 5 per cent solution of acetic acid stop bath for 15 to 20 sec. At this point, you may turn on room lights.

Discard developer A as soon as it becomes dirty or is greatly reduced in level by evaporation. Discard activator B daily, or when it has developed 50 sq ft of film per gallon.

Step 3—White-light Processing. White-light processing is accomplished in five steps, as follows.

Washing Out the Stencil. Remove unexposed, unhardened gelatin from the film by spraying it with water at 115 to 125°F. Place the film, emulsion up, on a drainboard or the clean back of an upturned tray, and let water from the spray fall directly on the emulsion. When the film is clear, it is ready for application to a screen. If the image washes away from the base, it indicates underexposure.

Preparation of the Silk. The screen on which to place the stencil must be absolutely clean. The proper way to clean a new stencil, and the correct procedure for removing ink and gelatin from previously used stencils are explained under step 4 of the Ektagraph process.

Adhering the Stencil. Place the wet stencil, emulsion up, on a table. Carefully put the screen in contact with the stencil, and cover both with a sheet of newspaper or a paper towel.

Make a pad of wadded cloth and blot up excess water by wiping across the sheet of newspaper or paper towel. Use heavy pressure while wiping to make a good contact between stencil and screen. Use several applications of blotting newspaper or paper towels, making sure you wipe all areas to ensure good adhesion. A print roller may be used in place of a wadded cloth, provided it is perfectly round and sufficient pressure is used.

Drying. Dry the screen at room temperature. For faster drying, direct a fan so air blows across the front of the screen. Drying with a fan is usually accomplished in about 20 min under average temperature and humidity conditions.

Stripping Off the Base. When a stencil is thoroughly dried, strip off the base by carefully lifting one corner. Then, roll the base back and away from the stencil. Mask the edges of the stencil when the base has been stripped off. The stencil is now ready for printing.

4.4 Making the Frame

You can easily make the frame on which to mount screens or you can have it made by a carpenter to your specifications.

The frame should be made of lumber strips no smaller than 1 by 3 in., cut to the length and width desired. To make a frame yourself, miter the ends of the wooden strips, and join them together with nails. Corrugated nails, such as those used in frame making, provide the strongest bind.

When the frame is completed, mount the screen in three steps, as follows:

1. Before cutting the screen, which is usually manufactured in rolls, measure the fabric to right size by spreading it across the frame from outside edge to outside edge. Cut the screen so it covers the entire wooden area of the frame, that is, from the outside edge of each wooden segment to the outside edge of its opposite wooden segment.

2. Remove the measured screen from the frame. With strips of cloth ribbon or felt, make a cushion around the entire frame. This cushions the screen, which will be stretched over the frame, so the edges of the wood will not cut into the screen. Lap the cloth ribbon or felt over the frame's inside edges.

3. Lay the screen on top of the frame and begin to staple silk to frame. Always start the stapling in the center to prevent ripples. After stapling one side, and as you staple the other sides, stretch the screen so it is tense. This tension provides a better printing surface. Remember to work around the frame, stapling in such a manner as to keep the screen even.

part **II** | **TECHNIQUES OF CONTINUOUS-TONE REPRODUCTION**

5 | REPRODUCTION OF CONTINUOUS-TONE ORIGINALS IN THE CAMERA

A CONTINUOUS-TONE ORIGINAL is a photograph or a piece of original artwork possessing a full range of tones. A black-and-white photograph or drawing that contains various shades of gray tones as well as black and white is a continuous-tone original. A piece of colored material (a color original) possessing a full range of colored tones is also a continuous-tone original.

Industrial, commercial, and even studio photographers can increase their work loads tremendously if they are able to offer customers good photographic copies of continuous-tone originals. A huge demand exists for technicians who can perform this task, because, unfortunately, not many have taken the time nor made an effort to perfect their copying technique.

Elaborate, expensive equipment is not needed for continuous-tone copy work. Unlike the precise reproduction of drawings required by process copying, which calls for the use of precision equipment, continuous-tone copying is often done with modest equipment. Results, however, can be excellent.

An able technician using a view or press camera can competently accomplish continuous-tone copy work. The job is made easier, of course, if a precision camera, such as a graphic-arts reproduction camera, is employed. On the other hand, continuous-tone copying can be done with roll-film cameras as well. Techniques are the same no matter what camera is used.

Users of larger-size pieces of equipment will find they have a wide and varied selection of special sensitive materials at their disposal which permit high-quality results. Materials available for copying with roll-film cameras are limited.

Much of the information in this chapter depends for its understanding on material presented in Chap. 1. Reference is made to the latter for basic data on cameras, lighting, and exposure.

Six terms used in this chapter that should be defined immediately are as follows:

To copy: To reproduce photographically continuous-tone and line subjects, such as photographs, paintings, and other printed or written information.

Original: The material to be reproduced.

Copy negative: The negative obtained by copying a continuous-tone original.

Reproductions: Photographs or prints made from a copy negative.

Monochrome: Subject matter of a single shade, which may possess various tones. For example, black-and-white photographs and toned prints are referred to as being monochrome.

Color original: An original in color that possesses continuous tones. For example, color photographs and color renderings are referred to as being color originals.

5.1 The Copy Negative

The reproduction technician is often asked to make copy negatives of continuous-tone originals. These negatives are used to produce photographic copies of the original. Production of copy negatives is necessary when an original negative of an existing photograph is not available or has been damaged, when copies of an original piece of continuous-tone artwork are needed, or when retouching, such as airbrushing, has been done to a black-and-white continuous-tone original. By making a copy negative in the latter case, the technician can reproduce as many of the correct, retouched photographs as desired, and will also have available a correct, permanent record of the retouched original.

Since photographic prints are to be made from a copy negative, you must render *as faithfully as possible* the varying range of tones contained in the continuous-tone original you are copying. The copy negative of a continuous-tone original should produce photographic prints as good in quality as the original; in other cases, the copy negative may produce prints of better quality than the original; in no case should the copy negative ever produce prints inferior in quality to the original.

How do you know when you have obtained a good copy negative? Generally, three factors characterize a distinct copy negative. These are (1) higher overall density than that normally possessed by an original negative, (2) normal contrast, (3) good detail in shadow areas.

Higher overall density is achieved in the camera stage; contrast and good shadow detail are achieved in the camera and development stages.

Density. The reproduction quality of a copy negative depends directly on the overall density of that negative. *Generally, expose continuous-tone originals to obtain copy negatives of higher densities than are usually desired in*

FIG. 5.1 A continuous-tone original suitable for copying. The gray scale illustrates the various shades of gray tones of the original.

original negatives (Fig. 5.1 and 5.2). To know when proper density has been reached, you must rely on your own experience or use a densitometer.

When making prints from an *original* negative, it is possible to reproduce accurately the various tones with low or moderately low negative density. Many photographers, in fact, prefer to work with a low-density original negative, especially when enlarging. This degree of density, however, is *not* desirable in a *copy* negative, since low density produces continuous-tone prints in which white and light tones tend to darken.

Conversely, a copy negative of higher density permits you to retain true white and light shades of gray, without having to print on a hard grade of paper.

FIG. 5.2 A copy negative of the continuous-tone original shown as Fig. 5.1. Note the slightly higher than normal density.

Contrast and Detail. Contrast and detail in copy negatives are other important considerations for the reproduction technician.

Normal contrast in a copy negative, which is desirable, is often difficult to achieve when you are copying a continuous-tone original that is either flat (lacking contrast) or is too contrasty. Through the use of various techniques, such as manipulation of development time, and by employing suitable films and developers, you can, however, control contrast.

When copying continuous-tone originals of *normal* contrast, *always expose to the fullest and underdevelop slightly*. This procedure provides copy negatives of normal contrast and with a fine grain structure.

The remainder of this chapter discusses the three types of continuous-tone original copying jobs you may be asked to perform: (1) copying black-and-white continuous-tone originals in black and white; (2) copying color continuous-tone originals in black and white; (3) copying color continuous-tone originals in color.

5.2 Basic Data for Copying Black-and-white Material in Black and White

Orthochromatic films, commercial films, and panchromatic (type B) films can be used for making copy negatives from black-and-white continuous-tone originals.

Orthochromatic Films. These are sensitive to the blue and green regions of the spectrum, but have no response to the red region. Those areas of the film exposed to red do not record the image and develop clear. Because of this insensitivity, orthochromatic films can be handled safely during development under the subdued light of a red bulb or a white light that has a red filter over it.

Some orthochromatic films produce excessive contrast and are not suitable for making copy negatives from black-and-white continuous-tone originals. Specifically, three special orthochromatic materials are designed for this job: Eastman Kodak gravure, Gevaert Correctone (type 2), and Agfa LM. These products are ideal for solving many difficult copying problems and have made the reproduction of continuous-tone originals easier.

Kodak Gravure. This moderate-speed long-scale orthochromatic film, which provides increased highlight contrast, produces good-quality copy negatives from which distinct reproductions of black-and-white continuous-tone originals can be obtained (Fig. 5.3).

The exposure level of gravure film determines overall and highlight contrast of the negative; hence, use particular care when copying with this film. Do not underexpose, or the film's flexibility is lost.

During development, gravure film can be handled under a red safelight filter, such as the Wratten Series 1, in a proper safelight lamp equipped with a 15-watt bulb. Keep the film at least 4 ft from the lamp during development. For best results, develop in a tray of Kodak D:11 at 68°F for 6 min. If you use a tank, extend development to 8 min.

Gevaert Correctone (Type 2) and Agfa LM. These two orthochromatic films are unusual and versatile materials, which can be employed for either single exposures or double exposures. They are handled in the same manner and produce similar results.

As materials designed for dual exposures, Gevaert Correctone (type 2) and Agfa LM can be used to copy originals that possess both black-and-white continuous-tone areas and line work. They can also be employed for dropping out the white backgrounds of wash drawings, exploded views, and other types of artwork.

Each film is composed of two separate and distinct emulsion layers: a top emulsion that is composed of a high-contrast orthochromatic layer, and an underside emulsion that is composed of a long-scale color-blind layer. Consequently, you must use yellow and light blue filters when making dual exposures. The yellow filter prevents light from exposing the high-contrast orthochromatic emulsion while the underside color-blind emulsion is being exposed. Similarly, the blue filter prevents exposure of the long-scale color-blind emulsion while the high-contrast orthochromatic emulsion is being exposed.

Expose the continuous-tone image of the original first. This is recorded on the long-scale color-blind emulsion and is an exposure of shorter duration than the second. During the second exposure, the line or high-contrast copy is recorded on the high-contrast orthochromatic emulsion.

The versatility of Gevaert Correctone (type 2) and Agfa LM enables you to produce excellent copies of combination originals without having

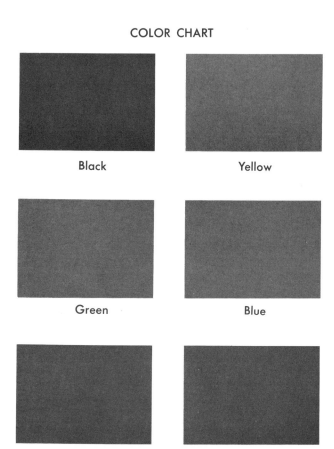

COLOR CHART

Black

Yellow

Green

Blue

Orange

Red

FIG. 5.3   A color chart copied on Kodak gravure film.

COLOR CHART

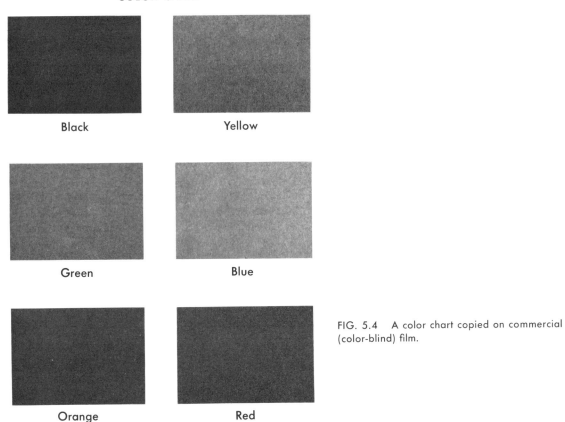

Black   Yellow

Green   Blue

Orange   Red

FIG. 5.4 A color chart copied on commercial (color-blind) film.

to make two separate negatives—one for the continuous-tone image and one for the line material—and stripping them together. It also permits you to make combination copy negatives in those cases where stripping is not possible.

Gevaert Correctone (type 2) and Agfa LM can be exposed by tungsten bulbs, arc lamps, or Day-Lite fluorescent tubes. Use Dektol, Hunt-7, or DK:50 for development.

Commercial (Color-blind) Films. These are specifically manufactured for commercial processing. They are sensitive to the violet and blue regions of the spectrum only, and do not record green, yellow, and red in their true values (Fig. 5.4). Because of their lack of response to red, you can safely handle color-blind materials under the subdued light of a red bulb or a white light that has a red filter over it.

Commercial films produce negatives of medium-to-high contrast. They are particularly suited for making copy negatives of original artwork, such as wash drawings and other black-and-white renderings not possessing a complete scale of tones. With careful handling, however, they also provide good results when reproducing other types of black-and-white continuous-tone originals that have a full range of tones, such as photographs.

Develop commercial film in a DK:50-type developer, either full strength or diluted 1:1 for softer results. Remember: this film can produce excessive contrast; so use caution in developing. Never overdevelop or you will get copy negatives possessing highlights that are difficult to print.

Commercial films come in acetate and matte bases that can be retouched without the use of retouching fluids, making it possible to apply ink or pencil retouching agents directly to the film without using the retouching fluids normally required by other films.

Panchromatic (Type B) Films. These are used only in emergencies when films more suitable for copying black-and-white continuous-tone originals are not available. An excellent conventional photographic medium, which also produces good black-and-white copy negatives from *color* continuous-tone originals, this film does not have the flexibility needed to make suitable copy negatives of black-and-white continuous-tone originals (Fig. 5.5).

Being sensitive to all colors (particularly to green), panchromatic (type B) films must be developed in total darkness. Hence, control over development is held to a minimum, since you cannot inspect the film under a red safelight. Another disadvantage is that panchromatic (type B) films do not record highlights as faithfully as orthochromatic and commercial films.

There is one condition under which you have no alternative but to employ panchromatic film: when copying black-and-white continuous-tone originals with a roll-film camera. Orthochromatic and commercial materials suitable for continuous-tone copying are not available in rolls.

When this situation arises, use a medium slow-speed, fine-grain emulsion

COLOR CHART

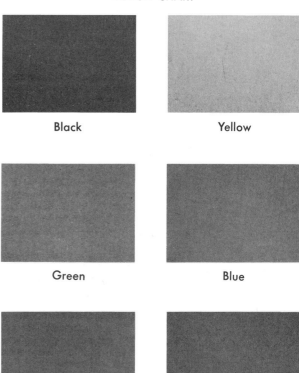

Black

Yellow

Green

Blue

FIG. 5.5   A color chart copied on panchromatic (type B) film.

Orange

Red

film, such as Kodak's Plus-X or Panatomic-X, Ansco's All-Weather Pan, or Gevaert's XL Pan. Although slow in speed when employed under tungsten illumination, these films offer high resolution and extremely fine-grain characteristics. They can provide negatives from which you can make adequate enlargements, provided exposure and development operations are done with extreme care and accuracy.

Never use a high-speed roll film when making copy negatives from black-and-white continuous-tone originals. These so-called available light materials usually produce excessive grain, resulting in copy negatives of poor quality.

5.3 Basic Data for Copying Color Material in Black and White

A basic knowledge of color sensitivity, filters, and suitable sensitive materials is necessary for reproducing color originals in black and white. Color originals take such form as continuous-tone and halftone prints, original paintings, transparencies, and wash drawings. Being able to evaluate an original and plan your approach is necessary before you can properly record color values in their true relationships.

Light, like sound, is composed of a series of waves. Long wavelengths produce the sensation we see as red, medium wavelengths produce green, and short wavelengths produce blue. A combination of all wavelengths, in proper balance, produces white light (Fig. 5.6). Thus, if you analyze white light under a spectroscope, you find it consists of these three primary colors—red, green, and blue. Regions between primary colors consist of combinations of these colors called *secondary colors* (orange, yellow, and violet are a few examples). All colors, whether primary or secondary, are arranged in bands according to their wavelengths.

Objects appear to be of different color because they absorb certain wavelengths from the spectrum and reflect others. In copying a color original, you must consider it as subject matter that not only reflects the colors you see but also reflects and absorbs other colors that you do not see.

One of the most difficult problems in copying color originals is to reproduce colors in their true relationships. A color rendering, for example, possesses various visible colors and shades of colors that have a definite contrast and separation. In reproducing these in black and white, care must be taken to retain the color separation portrayed in the original. To accomplish this, a reproduction technician must know the best film to use. He must, in addition, have knowledge of filters, for with these he can increase, decrease, or maintain on the reproduction the color values of the original in their true balance, or he can exaggerate the effect to suit a particular requirement (Fig. 5.7).

Films for Copying Color Originals in Black and White. The emulsions of all films are sensitive to one or more colors and insensitive to others. The reproduction technician must know the relationship between the films he uses and their color sensitivity.

The following films can be used for copying color originals in black and white.

Panchromatic Films. Panchromatic films are sensitive to practically all

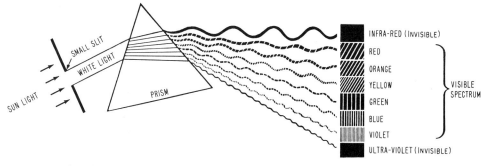

FIG. 5.6   The spectrum as produced by passing white light or sunlight through a prism.

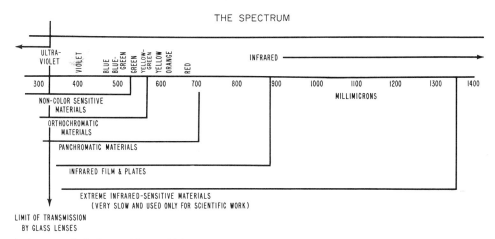

FIG. 5.7   The spectral range of photographic sensitivity.

colors and produce monochrome results that approximate what the eye sees. They are the best materials to use for copying color originals in black and white.

There are three general types of panchromatic emulsions, as follows:

1. Type A emulsions are process-type materials that record red darker and blue lighter than their true values. Use this film when desiring contrast effects. For less contrasty results when using tungsten illumination, employ a K-2 yellow filter, which lightens red areas and darkens blue areas.

2. Type B emulsions are considered the best to use for copying color originals in black and white. They provide higher speed and better color correction than other panchromatic emulsions. When using incandescent illumination, type B emulsions give excellent renditions, even without employing filtration.

3. Type C emulsions, generally, are not recommended for copy work. When copying color originals under incandescent illumination, type C emulsions are too sensitive to the red region of the spectrum. If you must use this emulsion, employ an X-1 green filter to help correct red sensitivity.

Commercial (Color-blind) Films. As discussed in Sec. 5.2, commercial (color-blind) films are used primarily for reproducing black-and-white continuous-tone originals. If necessary, they may be employed in a limited way for copying color originals in black and white, since they are sensitive only to the blue-violet band of the spectrum. They are ineffective for recording the other colors that may be employed in a color original.

Color-blind films record blue as light, red as black, and other colors much darker than their true tones. If you desire to lighten blue and darken red on a copy negative, the color-blind emulsion can be used instead of a panchromatic emulsion and a blue filter which produce the same effect.

Orthochromatic Films. As with commercial films, orthochromatic materials are used only for special color original copying jobs. Orthochromatic films are sensitive to blue and green, but insensitive to red. Thus, you may use this film to copy color originals on black-and-white copy negatives if retention of red in its true value is not particularly important. Similarly, orthochromatic films may be used without a filter when it is necessary to darken red on a reproduction.

Theory and Application of Filtration. Filters are employed to achieve one or a combination of the following purposes.

Contrast. Some filters record originals on black-and-white negative material in unnatural, sharply contrasting tones. These filters are often used to produce artistic effects or for special purposes.

Correction. Filters are often used for reproducing color originals in monochromatic tones, which appear natural to the eye on a black-and-white copy negative.

Separation. Filters are often used preparatory to color printing. They render colors of original material in separate tones; so these colors can be registered individually on separation negatives.

Special Purposes. Reproduction technicians often use special filters (such as neutral density and polarizing) to cope with difficult lighting conditions.

In Chap. 1, a basic rule governing the use of filters was presented: each of the three primary types of filters—red, green, and blue—tends to lighten its own color and darken the other two. To these should be added another important filter used in reproduction—the yellow filter. Yellow lightens its own color and darkens blue.

Filters either absorb or reflect light. Light that is not absorbed by a filter is transmitted. Transmitted light acts on a film's emulsion. Thus, when using filters, your main concern is light that is transmitted (Fig. 5.8).

Often, in copying of color originals, the light recorded on a black-and-white film is a combination of colors, such as blue-green and red-orange. If you wish to use a filter, *you must select one that transmits or holds back the*

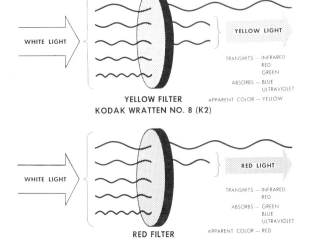

FIG. 5.8 The light-transmission principle upon which color filters work. (*Courtesy of Eastman Kodak Company.*)

Color of Filter	Effects on the Negative	
	Light held back (will print darker)	Light transmitted (will print lighter)
Yellow.............	Blue	Red and green
Green.............	Blue and red	Green
Red.............	Blue and green	Red
Blue.............	Red and green	Blue

FIG. 5.9   Effects of filters on the negative.

greatest amount of the color you wish to correct. Figure 5.9 will assist in making this selection.

These values are not absolute. For example, a green filter does not absorb all the blue or all the red, nor does it transmit all the green. The same applies to other filters for their respective colors. Thus, if you are using a filter to lighten or darken a color, and it is not doing an adequate job, substitute a lighter or darker filter, as required.

Most filters absorb a portion of all light reaching them, letting only a percentage of light through to the film. For this reason, an increase in exposure, called the *filter factor,* is required.

The filter factor varies with the type of illumination employed, since the composition of different lighting systems varies. The filter factor for Kodak gravure film using a K-2 yellow filter with white flame arcs, for example, is 5. With other tungsten illumination, such as photofloods, the filter factor is only 2.5. All film packages contain filter factor information.

Colors copied on a black-and-white film often appear different in their tonal relationships than the colors of the original itself. You must, therefore, consider the type of filtration that assists in producing well-separated shades of gray. To use filters effectively, then, a technician must determine how a subject will appear on the type of film he is using. This requires familiarity with the color sensitivity of the film.

In this respect, you can use a simple procedure to provide a ready reference for filtering. Collect the data sheets of the films you normally use, and record details of their spectral response (color sensitivity) in tabular form. Mount this tabulation near your copying setup for reference while copying.

5.4   Techniques of Copying Black-and-white Material in Black and White, and Color Material in Black and White

The information in this section applies to copying both black-and-white and color originals in black and white. It is divided into specific applications and problem areas to assist you in performing continuous-tone copying.

Determining Exposure. To establish the most accurate exposure levels for the films you employ in continuous-tone copying, make a series of exposure tests for each film, including a gray card scale test. Using meter readings and bellows factor determinations as a starting point (Sec. 1.2), bracket ex-

DESCRIPTIONS AND USES OF FILTERS

	Number	Description and Use
COLORLESS	0	For compensating thickness of other gelatin filters in optical systems
	1A	Kodak Skylight Filter—Reduces excess bluishness in outdoor color photographs in open shade under a clear, blue sky
YELLOWS	2A	Absorbs radiant energy below 405 mμ
	2B	Absorbs ultraviolet below 390 mμ
	2C	Absorbs ultraviolet below 385 mμ
	2E	Absorbs radiant energy below 415 mμ
	3	Light yellow (Aero No. 1)
	3N5	No. 3 plus 0.5 neutral density
	4	Yellow—Approximate correction on panchromatic materials for outdoor scenes, including sky
	6	K1—Light yellow—Partial correction outdoors
	8	K2—Yellow—Full correction outdoors on Type B panchromatic materials. Widely used for proper sky, cloud, and foliage rendering. Green separation for Kodak Fluorescence Process
	8N5	No. 8 plus 0.5 neutral density
	9	K3—Deep yellow. Moderate contrast in outdoor photography (with black-and-white films)
	11	X1—Yellowish-green. Correction for tungsten light on Type B panchromatic materials; also for daylight correction with Type C panchromatic materials in making outdoor portraits, darkening skies, or lightening foliage
	12	Minus blue. Haze cutting in aerial photography
	13	X2—Dark yellowish-green. Correction for Type C panchromatic materials in tungsten light
	15	G—Deep yellow. Overcorrection in landscape photography. Contrast control in copying and in aerial infrared photography.
	16	Blue absorption
	18A	Transmits ultraviolet and infrared only (glass)
ORANGES AND REDS	21	Blue and blue-green absorption
	22	Yellow-orange. For increasing contrast in blue preparations in microscopy. Mercury yellow
	23A	Light red. Two-color projection—contrast effects
	24	Red for two-color photography (daylight or tungsten). White-flame-arc tricolor projection
	25	A—Tricolor red for direct color separation. Contrast effects in commercial photography and in outdoor scenes. Two color general viewing. Aerial infrared photography and haze cutting
	26	Stereo red
	29	F—Deep red. Red color separation from transparencies and for the Kodak Fluorescence Process. Strong contrast effects. Copying blueprints. Tungsten tricolor projection
	92	Red. For densitometric measurement of color films and papers
MAGENTAS AND VIOLETS	30	Green absorption
	31	Green absorption
	32	Minus green
	33	Strong green absorption
	34	Violet
	34A	Blue separation—Kodak Fluorescence Process
	35	Contrast in microscopy
	36	Dark violet

FIG. 5.10   The filters described here are the Eastman Kodak Wratten filters, which are the ones most commonly used in conventional photographic and scientific work. (*Courtesy of Eastman Kodak Company.*)

DESCRIPTIONS AND USES OF FILTERS

	Number	Description and Use
BLUES AND BLUE-GREENS	38	Red absorption
	38A	Red absorption. Increasing contrast in visual microscopy
	39	Contrast control in printing motion-picture duplicates (glass)
	40	Green for two-color photography (tungsten)
	44	Minus red—Two-color general viewing
	44A	Minus red
	45	Contrast in microscopy
	45A	Blue-green. Highest resolving power in visual microscopy
	46	Blue projection (experimental)
	47	C5—Tricolor blue for direct color separation and from Kodak Ektacolor Film for Dye Transfer. Contrast effects in commercial photography. Tungsten and white-flame-arc tricolor projection
	47B	Tricolor blue for direct color separation, separation from transparencies and from Kodak Ektacolor Film for graphic arts
	48	Green and red absorption
	48A	Green and red absorption
	49	C4—Dark blue
	49B	Dark blue
	50	Very dark blue. Mercury violet
	94	Blue. For densitometric measurement of color films and papers
	98	Blue. For use with Kodak Ektacolor Paper—equivalent to a No. 47B plus a 2B filter
GREENS	52	Light green
	53	Medium green
	54	Very dark green
	55	Stereo green
	56	Very light green
	57	Green for two-color photography (daylight)
	57A	Light green
	58	B—Tricolor green for direct color separation. Contrast effects in commercial photography and microscopy
	59	Green for tricolor projection (white-flame-arc)
	59A	Very light green
	60	Green for two-color photography (tungsten)
	61	N—Green color separation from transparencies and Kodak Ektacolor Film. Tricolor projection (tungsten)
	64	Red absorption (light)
	65	Red absorption
	65A	Red absorption
	66	Light green. Contrast effects in microscopy and medical photography
	93	Green. For densitometric measurement of color films and papers
	99	Green. For use with Kodak Ektacolor Paper—equivalent to a No. 61 plus a No. 16 filter
NARROW-BAND	70	Dark red. For use with Kodak Ektacolor Paper and color separation from Kodak Ektacolor Film (with tungsten)
	72B	Dark orange-yellow
	73	Dark yellow-green
	74	Dark green. Mercury green
	75	Dark blue-green
	76	Dark violet (compound filter)
	77	Transmits 546 mμ mercury line (glass plus gelatin)
	77A	Transmits 546 mμ mercury line (glass plus gelatin)

FIG. 5.10 (continued)

DESCRIPTIONS AND USES OF FILTERS

	Number	Description and Use
PHOTOMETRICS	78	Bluish. Photometric filter (visual)
	78AA	Bluish. Photometric filter (visual)
	78A	Bluish. Photometric filter (visual)
	78B	Bluish. Photometric filter (visual)
	78C	Bluish. Photometric filter (visual)
	86	Yellowish. Photometric filter (visual)
	86A	Yellowish. Photometric filter (visual)
	86B	Yellowish. Photometric filter (visual)
	86C	Yellowish. Photometric filter (visual)
LIGHT BALANCING	80B	Blue. Kodak Photoflood Filter for Kodak Daylight Type Color Films
	80C	Blue. Kodak Photoflash Filter for Kodak Daylight Type Color Films
	81	Yellowish. For warmer color rendering
	81A	Yellowish. For Kodak Ektachrome Film, Type B, with photographic flood lamps
	81B	Yellowish. For warmer color rendering
	81C	Yellowish. For Kodachrome Film, Type A, with flash lamps
	81D	Yellowish. For Kodachrome Film, Type A, with flash lamps
	81EF	Yellowish. For Kodak Ektachrome Film, Type B, with flash lamps
	82	Bluish. For cooler color rendering
	82A	Bluish. For Kodachrome Film, Type A, with 3200 K lamps
	82B	Bluish. For cooler color rendering
	82C	Bluish. For cooler color rendering
	83	Yellowish. Kodachrome Commercial Film filter
	85	Orange. Kodak Daylight Filter for Kodak Type A Color Films
	85B	Orange. Kodak Daylight Filter for Kodak Type B Color Films
	85C	Light orange. Kodak Daylight Filter for Kodak Type F Color Films
	85N3	For daylight exposure of Eastman Color Negative Film at relatively large apertures
	85N6	For daylight exposure of Eastman Color Negative Film at relatively large apertures
MISCELLANEOUS	79	Photographic sensitometry. Corrects 2360 K to 5500 K
	87	For infrared photography. Absorbs visual
	87C	Absorbs visual, transmits infrared
	88A	For infrared photography. Absorbs visual
	89B	For infrared photography
	90	Monochromatic Viewing Filter
	96	Neutral filters for controlling luminance
	97	Dichroic absorption
	102	Conversion filter for Barrier-layer photocell (to luminosity)
	106	Conversion filter for S-4 type photocell (to luminosity)

FIG. 5.10 (*continued*)   The filters described here are the Eastman Kodak Wratten filters, which are the ones most commonly used in conventional photographic and scientific work. (*Courtesy of Eastman Kodak Company.*)

posures so normal, underexposed, and overexposed negatives are produced. Develop all films in the same developer, at the same temperature, and for the same period.

When negatives are dry, make a good contact print of each on a normal grade of paper, without using printing techniques (burning-in, dodging, etc.). Match these prints against the continuous-tone original, and determine which one is closest in quality to the original. Note the exposure used. This exposure now serves as the basis for all future work done with that type of film.

When making prints from a copy negative at this exposure, you can obtain more contrast by printing the negative on a harder grade of paper or by using a more active developing solution. Conversely, when desiring softer, less contrasty prints from a copy negative, print the negative on a softer grade of paper, use a more diluted developing solution, or use a soft-working developer.

If possible, test exposures should be made at a ratio of 1:1 (same-size) with the continuous-tone original, permitting you to calculate an exposure starting point and to determine accurately exposures for reductions in copy.

Lighting. Lighting equipment and techniques for continuous-tone copying are the same as those required for reproducing line material in the camera (Sec. 1.2).

Occasionally, lighting may produce objectionable surface reflections on the continuous-tone original being copied. This often happens when copying oil paintings, artwork that has been lacquered, and subject matter presented on an unusual surface, such as paintings and drawings rendered on metal, glass, or polished wood.

Reflections mar reproduction quality and should be eliminated. When using a view, press, or roll-film camera, eliminating reflections should be an easy task. Simply rearrange light sources. If this procedure does not work or if you are using a copy camera that has a fixed lighting arrangement, some other method for reducing reflections has to be found.

Some technicians often try the *submergence technique* to minimize reflection. They submerge an original in a tray of water, image up, and photograph it through the water. This method may or may not work, and it can only be used if the surface of the original being copied has a rough texture.

Another technique you can try is the *glycerin method.* If reproducing a small-size photograph which has a damaged or cracked rough-textured surface that reflects light, soak the print in a solution of concentrated glycerin for a few minutes, remove it, and place it face down on a sheet of clear, clean glass. Then, squeegee the print and photograph it through the glass. When the job is completed, wash the print in water to remove all glycerin. This method can only be used when copying black-and-white originals. It cannot be used for copying color originals, since color surfaces may be damaged by glycerin and rough handling.

The surest way to eliminate objectionable reflection is to polarize light entering the camera's lens by means of *polarizing filters.* These are neither custom nor expensive pieces of equipment, but are standard for professional laboratories.

There are two types of polarizing filters: one type is placed over the camera's lens; the other type, which can be used with the lens type, is placed over each light source (Fig. 5.11). With these aids, the reproduction technician has maximum control over his light sources and over the light entering the lens.

A polarizing filter used on a lens provides a technician with a great deal of latitude, but its corrective powers can be substantially increased if it is employed with polarizing filters on light sources. Only nonmetallic sur-

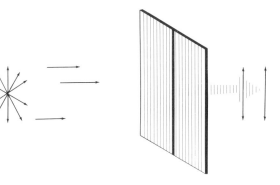

FIG. 5.11 The principle of polarizing filters. The arrows at the left represent light vibrating in many directions. As light hits the polarizing filters, rays not parallel to the filter's openings are blocked off, while those that are parallel are allowed to pass through. Light that passes through, represented by the arrows on the right, is plane-polarized light that vibrates in one direction only.

faces can be polarized by using a filter on a camera's lens. By adding light filters, you can increase the correction factor by being able to copy images made on metallic surfaces as well.

Ordinary light vibrates in all directions perpendicular to the light's ray. With a polarizing filter placed over each light source, the light's vibration path is controlled, making it a straight line, a circle, or an ellipse, depending on the type of filter employed. In copying, only that polarizing filter that produces a straight-line vibration (called *plane-polarized light*) is of interest. Polarizing filters that produce plane-polarized light limit the plane of light vibration to the lens, thereby reducing or eliminating entirely reflection and glare (Fig. 5.12).

Polarizing filters for lenses operate on a similar principle. When placed over a camera's lens, a filter is manipulated to admit light that vibrates in a limited plane only. To set a filter on the camera's lens in the correct position, hold it in front of your eye, and rotate it until glare and reflection emanating from the original are eliminated. Then, check the position of the filter arm, and set the filter into the lens holder on the camera in that position. The correct position can also be obtained by placing the filter in its holder on the camera and rotating it while checking effect through the camera's ground glass.

The filter factor using lens' polarizing filters with panchromatic film is 2; with commercial and orthochromatic films, the filter factor is 3.

No matter which method is used to reduce glare—polarizing filters, submergence, glycerin, adjusting lights—the easel to which an original for copying is attached should *always* have a black background. Backgrounds made of white matte paper or other light-color materials add to the reflection problem. You can make a suitable background with a sheet of black

FIG. 5.12 A typical copying set up using polarizing filters over each light source to provide maximum glare reduction. (*Courtesy of Eastman Kodak Company.*)

paper or by painting the easel with a flat black paint. Glare and reflection bounced by light shining on surrounding items, such as metal, will not be picked up by the easel area and reflected through the lens.

Transillumination. Problems often arise in reproducing shadow area details of a very dense black-and-white continuous-tone original. These details are often not distinguishable under conventional illumination and, consequently, tend to "blacken in" on the copy negative. Thus, transillumination is an invaluable technique to use for drawing out highlights of dense shadow areas.

Set up two light sources. One should be a light box or transparency viewer, which contains a sheet of ground or opal glass over the light, thus diffusing light and preventing hot spots on the copy negative. The original is taped to this, providing a rear-light that shines *through* the print. The other light source should be conventional illumination placed at a 45° angle to the copy setup.

Take a meter reading with only the front-light source turned on, and make a test exposure with this lighting for one-half the exposure calibrated on the meter. Now, turn off the front-light source and all room lights. Turn on the rear-light source, which shines through the print and increases shadow detail. Make a second exposure for one-half the exposure calibrated on the light meter, that is, for one-half the meter reading you determined for the front-light exposure. For example, if the meter reading obtained for the front-light source was 4 sec, make one exposure of 2 sec with the front lights turned on. Then, with the front lights turned off and the rear lights turned on, make another exposure of 2 sec.

After developing the test negative, determine whether shadow detail is adequately recorded. If not, make another test exposure reducing front-light exposure and increasing rear-light exposure. Once correct exposure is established, all transillumination exposures made when copying an original can be started from this point.

A transparency viewer is an excellent device to use as a transillumination rear-light source (Fig. 5.13). It is available in sizes large enough to accommodate original prints of 8- by 10-in. size. If a light table is available, it too can be effectively employed to provide rear illumination.

Never use the transillumination method if there is writing or other markings on the back of a print, since these could be picked up by the lens and recorded on the copy negative.

FIG. 5.13 A typical transillumination setup, using a transparency viewer as a light source.

Copying Color Transparencies. The *transillumination method* can be used for copying color transparencies on black-and-white film. *The entire exposure, however, is made by use of the rear-light source.* No front lighting is employed.

It is particularly important to employ diffused lighting for transparency copying. A panchromatic film provides the best copy negative, with Eastman Kodak Super XX Pan film being most suitable.

Tape the transparency flat against the front of a light box. Mask off, with black paper, undesirable areas, particularly those that permit light to show around edges of the transparency, increasing glare. If illumination is sufficient to make an exposure, obtain a meter reading from the lighted transparency. This is the basic exposure. If illumination passing through the transparency is insufficient, making it impossible to obtain a direct meter reading, remove the transparency, and take a reading directly from the light source. After retaping the transparency to the light box, make a basic exposure by doubling the meter reading obtained directly from the light source.

With this method and a good bellows camera, you can copy color slides as small in size as 35 mm and magnify the negative to twice the size of the original. This method is more effective than employing a 35-mm camera with extension tubes for copying 35-mm color slides to black and white.

Color transparencies may also be copied by the *enlarger technique,* but exposures must be made in the dark since panchromatic film is used. Place a transparency original in the enlarger's negative carrier, and align it properly by placing a film holder or easel containing a sheet of white paper under the lens. Turn on the enlarger so that the transparency image is projected onto the white paper. Bring the image into proper focus, and mask the spot where the film holder or easel is positioned. The mask can be made of masking tape or cardboard. It permits you to position properly a holder or easel containing film while working in the dark.

After masking off the designated area, place a film holder or an easel loaded with film in the spot under the enlarger outlined by the mask. Lights must be off, of course. If using a film holder, pull its slide, leaving the film exposed, and make an exposure by turning on the enlarger for the required time (Fig. 5.14). Put the slide back into the holder after exposure.

As you should do when employing any type of illuminating technique for the first time, make a number of tests, using the enlarger method with transparencies of different characteristics. After tests have been completed, your procedure can be standardized since you will have test negatives by which to determine exposure for each type of transparency.

A film such as Kodak Super XX is recommended for enlarger copying, although any medium-speed panchromatic film may be used.

When developing black-and-white copy negatives made from transparency originals, remember that a color medium composed of dye images has been converted to a black-and-white medium composed of silver particles which could produce graininess if not properly developed. A copy will never equal the quality of an original transparency, but care in exposure and development will produce copy negatives of reasonable quality.

Films used to convert color transparencies to black-and-white copy

FIG. 5.14 The film-holder–enlarger method of copying color transparencies.

negatives should be developed in a medium-contrast developer, such as D:76, DK:50, and Isodol, which are diluted 1:1 to produce maximum shadow and highlight detail, medium density, and fine grain.

Producing Copies of Exact Size. Reproduction technicians are often asked to make copies of black-and-white and color continuous-tone originals that are the same size as the original. This is a requirement common in industry, particularly in engineering companies where measuring is done directly from copy prints or where prints are used for scale reference.

To produce copies the exact size as the original, use a commercial film possessing a dimensionally stable support, such as polystyrene or polyester. After developing the negative, make prints on a stable-base material, such as Eastman Kodak Resisto (contact or projection speed) or Du Pont Cronopaque. These produce good prints with excellent dimensional stability.

Films and paper with dimensionally stable bases provide accurate, same-size copies, whereas conventional films and papers do not. The latter tend to expand and shrink during processing, making it difficult to hold size.

Continuous-tone and Line Combinations. An everyday task for reproduction technicians is the preparation of continuous-tone reproductions that have included on them text or numerical information for such purposes as identification, advertising, and cataloging. Often, the information must be placed directly on the continuous-tone image area. An organization chart is an example of a continuous-tone original possessing text and tone area (Fig. 5.15).

Care in pasting up line information in its proper position on the toned area preparatory to reproducing both as a negative is a requirement of this type of work. Information can be placed right on the continuous-tone original or, if the original cannot be handled, on an acetate overlay. Use rubber cement for pasting up text, but be sure that excess cement is removed after paste-ups have been positioned. Excess drops of cement may

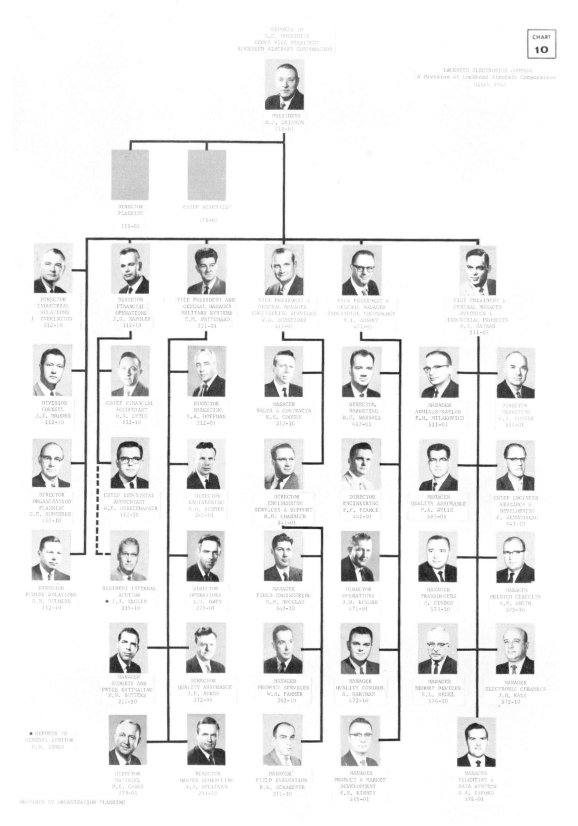

FIG. 5.15 A continuous-tone reproduction that includes text for identification purposes. (*Courtesy of Lockheed Electronics Company.*)

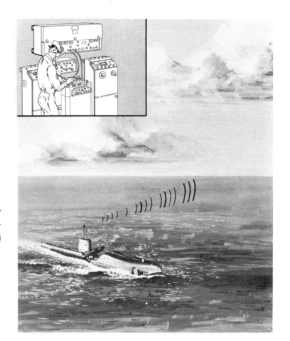

FIG. 5.16 A copy of an original possessing continuous-tone and line images rendered faithfully on Gevaert Correctone (type 2) film. (*Courtesy of Lockheed Electronics Company.*)

photograph on the copy negative and show up on the reproductions. To clean off rubber cement, use an artist's pickup or rubber cement pickup, both of which are available in art-supply stores.

When photographing combinations of line and continuous-tone material, correct exposure is of major importance if continuous-tone and line images are to be rendered satisfactorily in relation to one another. As explained in Sec. 5.2, Gevaert Correctone (type 2) and Agfa LM are specifically designed for this type of work (Fig. 5.16). Commercial (color-blind) and Kodak gravure films can also be used.

When employing commercial or gravure films, however, a compromise in exposure is usually necessary to balance the two contrasting types of subject matter. Generally, slight overexposure (about 20 per cent over normal) is desirable. Film should be developed to produce a slightly higher than normal contrast. Overexposure and slight overdevelopment tend to give better contrast between composite line and continuous-tone images (Fig. 5.17).

An alternate method of reproducing a combination original is by stripping together the line and continuous-tone images after making a negative of each. In other words, make a copy negative of the continuous-tone original and a line negative of the line material, thus providing better

FIG. 5.17 A copy reproduced on Eastman Kodak gravure film from a piece of continuous-tone artwork that possessed gradations of tones and solid blacks. Note the exceptional highlight quality. (*Courtesy of Lockheed Electronics Company.*)

FIG. 5.18 Inside of a typical light box. Note the wiring arrangement in the event you wish to build one. The box can be built to any size that serves your purpose. (*Courtesy of Eastman Kodak Company.*)

control over the reproduction of both types of images. *Caution:* Use films of the same thickness, so that there is no difference in the surface of the final negative after stripping. An irregular surface could prevent sharp prints, since negative-to-paper contact in the printer may be impaired.

One of two techniques can be employed to complete successfully the combination continuous-tone and line negative: the film-insert technique or the section technique.

Film-insert Technique. In this operation, sections of the continuous-tone negative are cut out, and sections of the line film are inserted correctly in these cutout areas. Work of this nature should be done on a light table or light box (Fig. 5.18). Proceed as follows:

1. Tape the continuous-tone negative to the light table or light box to prevent slipping. With the negative's emulsion side down, in contact with the light table's glass so that it will not be scratched, determine exactly where line inserts will be placed.

2. Make the cutouts with a new, sharp safety razor and a metal straightedge or T square. Be certain that alignment is accurate, or line inserts will be crooked when they are emplaced. Use a regular line-up table, if available, since the degree of alignment will be that much more accurate.

3. After openings have been cut, slip the line negative under the continuous-tone negative. Position the line information in its respective cutout opening (Fig. 5.19), and cut the line negative using the edges of the

FIG. 5.19 Film-insert technique. The reason the line copy is not visible in this photograph is because it is reversed to coincide with the reversal of the negative.

cutout openings in the continuous-tone negative as a guide. The end prod-uct is line film negative strips that are to be inserted by taping into the cutout openings of the continuous-tone negative.

4. Insert the line negative strips into their respective openings in the continuous-tone negative, and check positioning. Make sure the strips of line film are placed with their emulsion sides down to coincide for cor-rect reading with the continuous-tone negative, which also has its emulsion side down.

5. Tape the line negative strips into their respective cutouts in the continuous-tone negative with red or black graphic-arts cellophane tape, which is transparent and extremely thin. Line inserts should be taped flush with the surface of the continuous-tone negative. All taping must be done on the backside (nonemulsion side) of both films, so that tape does not interfere with negative-to-paper contact in the contact printer.

Taped areas appear as white borders on the reproduction. Thus, line negative inserts must be taped in some insignificant spot on the continuous-tone negative, such as in an irrelevant background area or on the bottom of the negative.

Section Technique. An entire section of a line film negative can often be added to the bottom of a continuous-tone negative and taped to it. This method, which is used primarily for adding titles, eliminates the cut-ting operation of the film-insert technique. When employing section stripping, make sure original line material is typed or drawn in its correct relation to the overall page size. Line material should never exceed the space available for it on the continuous-tone negative. If it does, the final continuous-tone–line negative may be too large for the contact printer.

1. Tape the continuous-tone negative to a light table, emulsion down. As was just mentioned, since you are adding a section of film to the continuous-tone negative, the final product could be a printing negative that is too long for your printing equipment. To eliminate this possibility, make the image area on the continuous-tone negative smaller, permitting the trimming off of excess border area and allowing the addition of line negative material without exceeding length requirements.

2. Measure the length of the line negative strip, and remove this amount from the continuous-tone negative by cutting with a sharp safety razor.

3. Butt the edges of line and continuous-tone negatives together (both with emulsion sides down) (Fig. 5.20). *Do not overlap them.* With the two flush, carefully tape them together.

FIG. 5.20   Section technique.

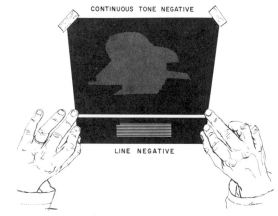

The finished product is a negative consisting of tone and line images, which is the correct size for printing on the required size paper in a contact printer or enlarger.

A word of caution is necessary when printing a combination line and continuous-tone negative. *Be sure to have good contact between film and paper.* A negative that has been spliced and taped must have excess pressure applied to it to guarantee absolute sharpness in the reproduction. A vacuum printer offers no difficulty, but those who use a conventional contact printer may have to add padding to the printer's cover to provide uniform pressure.

Occasionally, line inserts tend to be overexposed in printing, since clear areas of line material are exposed faster than continuous-tone images. If this happens, tape a few layers of wax paper or wrapping tissue over the line areas to even out exposure.

Line Masks. On occasion, you may have to copy a continuous-tone original, either a photograph or a piece of artwork, that has an objectionable background, such as extraneous objects or an uneven gray tone area that is undesirable. If the original is copied on a large-size film, such as 8 by 10 in., it is usually possible to opaque the negative, removing the background before producing prints. Opaquing smaller-size film, such as 4 by 5 in., however, is often a much more demanding and time-consuming job. In these cases, making a line mask with an orthochromatic process line film may solve the problem.

To employ the technique, place a clear acetate overlay over the original photograph or artwork. Then place a red, see-through acetate material, such as red Bourgess acetate or Ruby Studnite, over the overlay. With a sharp knife or razor blade cut the red acetate, tracing around the desired image area. *Remove that red acetate not covering the image.* The result is a red acetate mask over the desired image, with the objectionable background not covered by the red acetate.

Place a sheet of white paper under the red acetate overlay and make a line negative on a dimensionally stable lithographic film. Stable-base film permits accurate registration of the continuous-tone negative and line mask later on in the process. The developed negative possesses a black background and a clear image area. After the mask is developed, the acetate overlay can be removed and the continuous-tone negative can be made of the entire original.

The two negatives produced are the continuous-tone image negative, containing the image and the background, and the line mask, which possesses a black background to hold back the background on the reproduction and a clear image area to allow just the image to print through on the reproduction.

Now, tape the line negative mask over the developed continuous-tone copy negative, lining both up accurately. The combined negatives are then contact-printed or enlarged. The result is a print with the desired image area showing and a white background. All objectionable background areas are completely eliminated.

Typical Filtration Problems. The following are typical problems you may encounter in everyday routine that can often be solved by using filters.

FIG. 5.21   An example of what can be done with a faded photograph, yellowed by age and low in contrast. The print on the left was made on a portrait-type film; the photograph on the right was made on a contrast process orthochromatic film, using a light blue filter, and developed in a high-contrast developer. The correct combination of film, filter, and developer helped to produce a negative that provided a print with improved contrast and shadow rendition. (*Courtesy of Eastman Kodak Company.*)

Stained Prints. Photographic prints for copying may have colored stains present in the image area. A filter that is close in color to the stain will eliminate most of it in the reproduction. If the filter selected does not completely eliminate the stain on the reproduction, use one of a deeper shade. This technique is valuable when attempting to recover a seemingly ruined print.

Old, Faded Prints. Prints not too badly faded and with fairly good shadow detail can be copied on a commercial (color-blind) film, using a blue filter. If fading is extreme, the overall contrast of the original print may be low. Improve contrast and eliminate yellowish discoloration on the reproduction by using a contrast-process orthochromatic film and a blue filter. Develop until the desired contrast is obtained (Fig. 5.21).

Copying Color-printed Reproductions. If the dot structure of a colored lithograph (halftone) or any other colored reproduction produced by a printing process, is seen clearly when viewed in a camera's ground glass, make a copy on a panchromatic lithographic film, using suitable filtration (Sec. 5.3). If, however, the dot structure is of little consequence, make the copy as you would any continuous-tone original (Fig. 5.22).

Copying Toned Prints. When copying sepia or other brown-toned prints, use a yellow filter and panchromatic film, which reduces overall contrast.

Duplicating Original Negatives. Reproduction technicians are often required to provide duplicate (copy) negatives to a customer or to another company division located in some other part of the country. This work is as much a part of copying as is the photographing of original artwork to obtain a copy negative.

There are many methods to use for making duplicate negatives, such as

FIG. 5.22 A continuous-tone copy made from a full-color offset print. (*Courtesy of Eastman Kodak Company.*)

the diapositive and autopositive techniques. The procedure outlined here, however, is the easiest and the one which produces consistently good results.

1. Make sure the original negative is absolutely clean. Dirt, dust, water marks, fingerprints, and other foreign matter must be eliminated or they will show up on the duplicate negatives. To clean the original negative, apply film cleaner with a wad of cotton. If foreign matter persists, moisten (but do not make dripping wet) a wad of cotton with water and try to wipe marks off. As a last resort, rewash the negative.

2. Contact print the original negative onto a sheet of glossy photographic paper or a sheet of Eastman Kodak fine-grain positive film to obtain a film positive. If necessary, employ printing techniques (dodging, burning-in, or using a tissue-paper mask to hold back light) to produce a high-quality intermediate. The final duplicate negatives are made from this intermediate.

3. If the intermediate is on glossy paper, copy it photographically to make as many additional copy negatives as required. If the intermediate is on fine-grain positive film, contact print it on any negative material, such as commercial film, to produce the duplicate negatives. A fine-grain positive film intermediate can also be printed back on to fine-grain positive film to make as many duplicates as required.

Copying When Away from the Laboratory. A reproduction technician is not confined to his laboratory, especially when he specializes in the reproduction of continuous-tone material. His work takes him to many places, and he often has to manipulate in difficult, confined areas.

For example, you may be requested to copy artwork in an art gallery or museum, where subject matter to be photographed is located in an area that makes it impossible to employ conventional lighting. The situation may call for improvised camera techniques as well, because bulky equipment, such as large cameras and elaborate illumination sources, may not be permitted in the area. The use of smaller-size cameras, perhaps roll-film equipment, may be necessary. By following the techniques and suggestions outlined in this chapter for camera usage and the materials to employ, little difficulty should arise.

Lighting, however, is a different problem and is usually the most difficult. The following list describes some suggested methods that may be employed.

Available Light. Some areas have enough illumination to permit copying by available light exposure, but these instances are rare. If convinced that existing illumination is satisfactory, make exposures carefully. Use an exposure meter and, if necessary, filters. Be sure to allow for filter factors. Develop negatives for a contrast slightly higher than normal to compensate for the low contrast that is usually produced by available indoor illumination.

Flash Techniques. Flash techniques may consist of using two flashbulbs in extension guns or two electronic flash extension units to obtain as much even lighting as possible. Maximum flat lighting may be impossible, however, because work may be done in a very confined area, with light sources being emplaced, through necessity, close to the camera.

Naturally, try to place flash or electronic units at those points that give the best light coverage. It may be helpful to employ one or two large pieces of white cardboard as reflectors off which to bounce the light of *flash* units, thus providing even, soft illumination. Use diffusers over flash reflectors to reduce the possibility of harsh lighting when using direct flash.

Electronic flash is soft illumination and does not require diffusion, except when you desire to reduce the overall output of the light.

Tungsten Illumination. When studio-designed conventional lights can be used, the job of judging lighting arrangement is easier. It may still be necessary to compromise by setting lights in the most suitable location, although this location may not be the conventional one for copying. Greater control can be exercised over conventional light sources than any other, especially when employing polarizing filters for maximum correction of glare and reflection.

No matter what light source is used, compose subject matter on the camera's ground glass or in the view finder so the format is presented free from distortion. If employing a roll-film camera, carry a small carpenter's level to determine exact level of the camera. A level camera helps to overcome distortion that makes parallel lines of the original converge on the reproduction. If distortion cannot be corrected when copying, a partial correction can be made in the enlarger by the simple expedient of tilting the easel.

There are certain items of equipment a reproduction technician should carry when working away from the studio. Not all this equipment may be needed on any one job, but you should be prepared for any eventuality. A check list of this equipment is as follows:

1. Camera with adequate supply of film
2. Tripod
3. Carpenter's level
4. Steel tape measure
5. Exposure meter
6. Cable release
7. Filters for correction, and lens shade
8. Lighting equipment

   a. If flash: extra batteries, "synch" cord, extension guns, extension cords, diffusers, adequate supply of bulbs

   b. If tungsten: light stands, reflectors, spare bulbs, extension cords, polarizing filters for lights, diffusers for lights if polarizing filters are not available

9. Notebook to record exposures and development recommendations
10. White reflectors, if necessary
11. Extra lenses

Concerning the need for extra lenses, when working in confined areas that inhibit camera placement, a camera's lens coverage may be limited. It may not be possible, for example, to cover a large original with a lens of normal focal length, and a wide angle lens may be required.

In some cases—when using cameras that do not accept additional lenses, for example—it may be necessary to photograph the original in sections. Make sure the camera is level. Absolute accuracy of exposure and alignment is necessary to permit eventual splicing of separate prints. This is a difficult task, but once it is mastered, it can be used effectively with success.

5.5 Copying Color Originals in Color

Although the basic techniques of copying continuous-tone originals in black-and-white apply to copying color originals in color, the latter is a specialized field unto itself. What makes it so are the extra techniques and precautions one must take.

In copying in black and white, for example, an error in placing illumination sources would probably result in a certain amount of uneven density on the copy negative that may not be objectionable. When copying color originals with color films, however, the slightest error in placing lights would result in a change of color on the negative, which usually ruins the copy.

Color work is critical work, and the field is vast and continually changing. For these reasons, outlined here are only the most important data relative to color copying. It is impossible to go into great detail. Only a book devoted entirely to color work could hope to cover the field. The data presented in this section, however, plus information on copying in other sections of this chapter and in Chap. 1, will give you a starting point from which to begin color to color copying.

Illumination. Primary types of lighting systems suitable for copying with color film are as follows.

Controlled Light Sources. Controlled light sources permit regulation of the color temperature of lights to produce the exact temperature needed for balanced results. These sources include ColorTran and pulsed xenon lighting, both of which are discussed in Chap 1. Color temperature of a light denotes its color quality, which is measured in degrees Kelvin (°K). The more redness a light contains, the lower its degrees Kelvin; the bluer the light, the higher the Kelvin value. Thus, when we speak of color temperature, we are actually speaking of the Kelvin value of the light.

Floodlights. Floodlights can be divided into four basic types: photofloods (3400°K), reflector floods (3400°K), floodlamps (3200°K), and T-20 projection lamps (3200°K). If used carefully and properly, each type produces good results at a minimum investment.

Floodlights have a limited life and gradually change in color value as they are used. Exercise caution when using them, and compensate for the number of hours they have been employed. Information regarding compensation is supplied with these bulbs.

Every color film is balanced to produce excellent results when exposed to light of a specific color temperature. For best results, you must use color films with the light sources for which they are designed. If a particular film must be employed with light sources with which it is not compatible, a conversion filter can be used to change the color quality of the light so it coincides with that of the film. Always follow the recommendations supplied with the film regarding suitable lights and conversion filters.

All photographic lamps are designed to operate at a given voltage to produce proper color balance. A constant check must be kept on this voltage to ensure it meets color requirements. Do this by periodically checking electric circuits in a laboratory or studio to detect fluctuation from normal voltage output. It is important to remember that a decrease of 1 volt causes a 10° drop in Kelvin temperature, with a 2- or 3-volt decrease (or increase) being sufficient to completely ruin a copy.

Two things occur when voltage fluctuates: if voltage spurts higher than normal, bulbs burn at too high a color temperature, causing excessive blue in the color; if voltage drops lower than normal, the color produced is in the orange-red area.

If current fluctuates too much in your laboratory or studio (have an electrician check it), a positive control, such as that provided by a voltage regulator or Variac, may be needed to keep voltage at a constant level to copy lights. Investment in a control device will eventually save much time and money in remakes. Many voltage regulators on the market are specifically designed for photographic use.

Meters that record color temperature are also available. These permit a reading of a lamp's color temperature to be taken, and provide a basis for selecting correction and balancing filters. Meters, however, are expensive and are usually needed only for extremely critical color reproduction work.

As with black-and-white copying, you can employ polarizing filters over floodlights, lens, or both when copying with color film. These help to reduce unwanted reflection and glare. Employing the correct filter factor is essential, since polarizing filters greatly reduce light transmission.

Photoflash Lamps. Photoflash lamps are economical lighting mediums that produce uniform output from bulb to bulb. Some color films are balanced for use with clear bulbs, without the need of filters; other films can be used with these lamps, but need appropriate filtration. Film instructions in each package enumerate filtration requirements, if any.

Electronic Flash. The light given off from most electronic flash units is balanced for use with color daylight films, without the need of filters. In many cases, however, these units render color more accurately if used with daylight color film and a correction filter. Filter recommendations are supplied by the manufacturer and are usually packaged with the electronic flash unit.

Exposure. Until your color copying setup is standardized, make exposure tests using a color chart, which permits correct exposure determination of color reproductions.

Obtain a meter reading from an 18 per cent reflective gray card, and make three exposures: a normal exposure according to the meter, a second exposure one-half stop under normal, and a third exposure one-half stop over normal. When making exposures, all room lights must be turned off. Room light can disrupt color balance, especially if exposures are of long duration.

Color Film	Illumination			
	Electronic flash	Photoflood 3400°K lamps	3200°K lamps	Clear (white) wire-filled flash
Type A film:				
Kodachrome professional..	Not recommended	None	82A	81C
Tungsten type (B):				
Anscochrome (sheet).....	Not recommended	81A	None	81D
Super Anscochrome......	Not recommended	81A	None	81D
Ektachrome (B) E-3......	Not recommended	81A	None	81C
Ektacolor type (L).......	Not recommended	81A	None	Not recommended
Internegative*..........	Not recommended	Not recommended	CC-90Y	Not recommended
Type F film:				
Ektachrome...........	Not recommended	82A	82C	None
Kodachrome...........	Not recommended	82A	82C	None
Kodacolor..............	85	82A	82C	None
Ektacolor (S)...........	85	82A	82C	None
Daylight type:				
Anscochrome...........	81A	80B	80B + 82A	80C
Super Anscochrome......	81A	80B	80B + 82A	80C
Ektachrome E-2........	None	80B	80B + 82A	80C
Ektachrome E-3*........	None	80B	80B + 82A	80C
Kodachrome...........	None	80B	80B + 82A	80C

* 120/620 rolls, Ektachrome professional roll film and Kodak Internegative, also available in sheet sizes.

FIG. 5.23   Correct filters to use with various types of illumination sources suitable for copying color originals on to color film.

After color film has been processed, compare the three exposures with the color chart to determine which exposure is closest to the chart. No off-balance effects should appear if light sources are properly balanced for the film. If the reproduced color is not normal, however, the trouble may lie in voltage fluctuation, improper filtration, or improper light sources.

To obtain an accurate evaluation when checking test results against color charts, use an illuminator designed for viewing transparencies. Illuminators specifically balanced for color viewing are available at reasonable prices.

Negative and Positive Color Materials. *Negative color materials* such as Ektacolor, Kodak Internegative, and Kodacolor can be used for making color prints, color transparencies, and, in emergencies, black-and-white prints from the resulting color negatives.

If the end product is to be a color print, it would be advisable as well as economical to employ a negative material rather than a transparency stock. Sheet films available for professional color negative work are most practical for copy work, since they are more suited to the artificial illumination exposures found in copy setups. The only exception to this is the Kodak Internegative film, which is available in sheet or roll form.

One of the films suited for copy work is Ektacolor (type L), which is available in sheet film only. This film is designed for long exposures by 3200°K artificial illumination without filtration, but it can also be illuminated by 3400°K photofloods with an 81A filter.

Ektacolor (type L) must not be exposed for shorter than ⅕ sec nor for longer than 60 sec. If these exposure limitations are disregarded, the resulting color errors may be uncorrectable in the printing process. The effective speed of Ektacolor (type L) depends on the illumination level and exposure time. For example, if 3200°K artificial illumination (without a filter) is employed, the following indexes apply:

Exposure Time, Sec	Effective Speed
⅕	25/3°
1	20/2.5°
30	10/1.5°
60	10/1.5°

With 3400°K photofloods and an 81A filter, the exposure is 5 sec and the effective speed is 12/2°.

When using negative color film, a grey card should be included in the overall area but outside the usable image. The grey card, which should be 18 per cent neutral reflectance grey, becomes an important aid to the laboratory technician doing the color printing. When the negatives are read on a densitometer, the red, blue, and green densities of the grey card seen on the negative are compared with a known standard; hence, a filter pack and suitable exposure time required to make the color prints can be easily calculated.

Copy negatives can be made on Eastman Kodak Internegative film. Many laboratories are using this material because of its improved highlight and contrast control. With proper handling, this material will probably

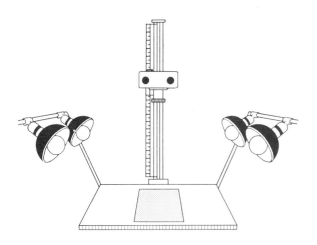

FIG. 5.24   A typical copying setup when employing 35-mm equipment for continuous-tone copying.

produce the maximum in color quality for copy work. Complete instructions concerning testing procedures for determining proper filtration and exposure are available with the film.

Positive color materials produce positive color transparencies directly, without an intermediate stage. They are suitable for viewing under transmitted light, and can also be used in the production of color prints and as a medium in making color-separation negatives for photomechanical printing processes.

Not as much control can be exercised over positive color material as over negative color film. It is easy to detect off-color results in transparencies when viewing them directly, but correcting the error in the printing process is difficult and often costly. With negative color film, color deviation is easily detectable when making a final print, and the situation can be corrected prior to production of the end product by filtration.

The types of color materials available for copying are limited in sheet sizes to Ektachrome and Anscochrome. Both are manufactured in a type B emulsion and are designed for use with 3200°K lamps, without filtration. Since no filter is needed, the films are easy to work with in copying.

Check emulsion numbers and special filter recommendations every time you open a new box of color sheet film, whether negative or positive. Since emulsions often vary from batch to batch, the manufacturer includes specific recommendations for use in each box. It is desirable to stock up with a supply of film having the same emulsion number for large-quantity jobs.

When using 35-mm cameras to copy originals for making color slides (Fig. 5.24), there are many types of films available. This number alone could be confusing. Avoid this confusion by buying and working with only one or two types.

Caution. Avoid the use of high-speed color film for copy work since these produce prominent grain structures.

Processing. Most reproduction technicians employ outside commercial processing laboratories specifically equipped for processing color film. Always be sure to include complete information with the exposed film, which should be carefully packaged. This information should include type of film, special instructions, if any, type of lighting employed, and ASA index used. This data will aid the processing laboratory and permit them to give the film the exact processing required.

6 | INFRARED, ULTRAVIOLET, AND RADIOGRAPHS

FOR NORMAL REPRODUCTION JOBS, the procedures discussed in Chaps. 1 and 5 on reproducing line and continuous-tone material, respectively, will permit the technician to handle most assignments. Unfortunately, reproduction is not always normal, and the technician could be presented with material for copying, which if reproduced by conventional methods and products would result in a reproduction other than that desired, to say the least. Such material includes drawings, documents, photographs, and so forth, that are damaged and even burned to an extent where all or part of the image area cannot be distinguished by the human eye.

Two processes that have gained in importance in recent years hold the key to reproduction of line and continuous-tone material considered unreproducible. These processes are infrared and ultraviolet photography, which have widened the scope of reproduction technicians by enabling them to complete successfully jobs considered practically impossible.

The invisible infrared band of the spectrum is light beyond the visible red band, while the ultraviolet band is invisible light below the visible violet band. Infrared is light at the extreme upper limit of the spectrum, while ultraviolet is light at the extreme lower limit of the spectrum.

Reproduction technicians once considered infrared and ultraviolet photography as scientific phenomena of a nature too technical for their use. It was not until the 1930s, for example, that those photographers who ventured into this area could actually purchase infrared film. Before this, they had to make their own film by a long, involved process that culminated in sensitizing a conventional film's emulsion with special dyes. During the 1930s, dyes were developed that permitted film manufacturers to easily

119

and economically make available films with emulsions already sensitized to infrared radiation.

This chapter introduces the subjects of infrared and ultraviolet photography. It should be emphasized immediately that if one of the processes does not produce the desired results, the other should be tried. No rules regarding the use of one or the other under particular circumstances can be expounded. In many cases, one method will work where the other will not. If both fail, however, the material to be reproduced is simply not reproducible.

Infrared and ultraviolet photography are used to reproduce originals of both a line and continuous-tone nature which cannot be satisfactorily reproduced by conventional methods. However, they are considered as continuous-tone reproduction methods since they employ films of a continuous-tone nature.

Section 6.3 is devoted to reproduction of radiographs (X rays), which are continuous-tone in nature. Frequently, the person producing the radiograph may desire additional copies, either in the form of film negatives or paper prints. The reproductions may be needed for distribution, as in industry where radiographs of metal castings are made, for submission to publications to accompany articles, such as unique medical X rays, or for filing. Reproduction of radiographs is an easy, but specialized technique, since one cannot simply place an X ray in a printer or enlarger and make a print from it, as may be done with other types of negatives.

Before beginning our discussion of infrared, ultraviolet, and radiographs, a word of caution is offered. All three fields are vast and continually growing. This chapter provides a practical introduction to all three. You will actually be able to employ any of the methods by putting into practice the information presented here. However, for refinement of the methods and for application in unique situations, books devoted in their entirety to the subjects should be consulted. A suggested list of these books is provided at the end of this book (see Additional Works of Reference).

6.1 Reproduction by Infrared Photography

Infrared radiation produces light waves which the human eye cannot see (Fig. 6.1). The eye has a spectral response of from 400 to 700 millimicrons. The infrared region of the spectrum begins at over 700 millimicrons and ranges upward to 1,300 millimicrons.

A millimicron, or one-thousandth of a micron, is the term applied in measuring wavelengths of light. A micron, in turn, is one-millionth of a meter. For our purposes, it is not important to understand how wavelengths are measured. It is important, however, to know that the human eye is sensitive to light—it sees light—only in the 400- to 700-millimicron range.

Infrared photography permits us to utilize light waves in the infrared region of the spectrum on materials sensitive to infrared light. Advances in perfecting dyes for sensitizing photographic materials now make it possible to record infrared light having a response as high as 1,300 millimicrons.

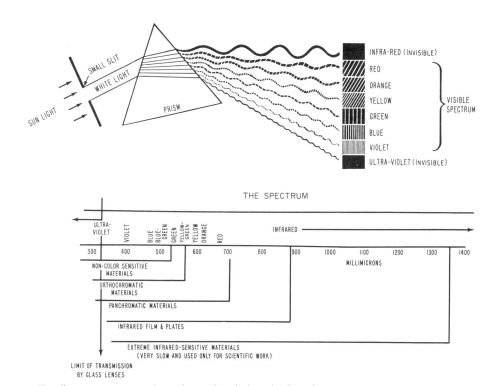

FIG. 6.1 The spectrum. The illustration on top shows how white light is broken down when passed through a prism. Note the relative positions of infrared and ultraviolet light. The illustration on the bottom shows the spectral range of photographic sensitivity, measured in millimicrons.

By using these sensitive materials, the photographer can obtain an almost perfect record of an image which has been damaged by time or accident or which has been altered (Figs. 6.2 to 6.5). This record often reveals portions of the original being copied that the photographer cannot see, but which the film can "see." Or it may reveal that the original contains different types of inks, pigments, carbon deposits, or other so-called foreign material which are identical in appearance to the eye. For this reason, infrared photography is used a great deal in criminology and by art appraisers and historians.

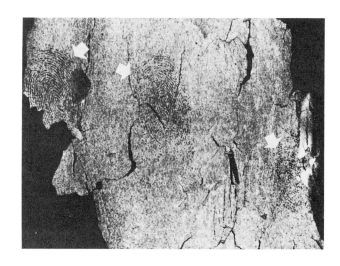

FIG. 6.2 Infrared reveals fingerprints. This area was photographed on infrared film through a Wratten Series No. 25 filter, with extremely flat lighting. Note the three fingerprints (arrows) that were brought out. (*Courtesy of Eastman Kodak Company.*)

UNITED STATES DEPARTMENT OF JUSTICE
FEDERAL BUREAU OF INVESTIGATION
WASHINGTON 25, D. C.

OFFICIAL BUSINESS

FIG. 6.3 Infrared penetrates ink. The dye-type ink used to obliterate an address (*top*) is penetrated by infrared photography to reveal the address (*bottom*). (*Courtesy of the Federal Bureau of Investigation, U.S. Department of Justice.*)

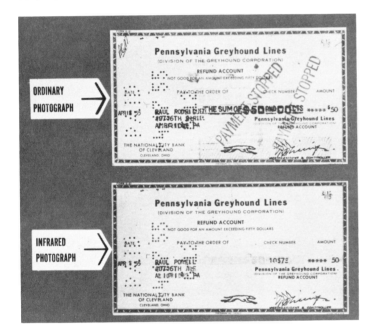

ORDINARY PHOTOGRAPH →

INFRARED PHOTOGRAPH →

FIG. 6.4 Check alteration found by infrared. The ordinary photograph (*top*) showed the check amount as $50. An infrared reproduction (*bottom*) revealed that the actual check amount was 50 cents. (*Courtesy of the Federal Bureau of Investigation, U.S. Department of Justice.*)

FIG. 6.5 Infrared brings out writing. These charred pieces of paper were worthless until photographed by infrared techniques, which brought out writing that had been obliterated. (*Courtesy of the Federal Bureau of Investigation, U.S. Department of Justice.*)

FIGS. 6.6 and 6.7   Greater detail through infrared. Figure 6.6 is a reproduction of an original painting. Figure 6.7 shows the same painting photographed by infrared. Note how infrared penetrated the surface layer of the painting to bring out subsurface detail. (*Courtesy of the National Gallery, London, England, and Eastman Kodak Company.*)

Suppose an old document has been altered by various ink compositions. These inks absorb infrared light in varied quantities, depending on the inks' chemical compositions and ages. These differences are usually not detectable by eye, and the inks may appear identical. When photographed with infrared materials and the proper filters, however, the differences become readily apparent, and each ink appears on the reproduction as having a different tone.

Suppose, also, that erasures (undetectable by eye) were made on this same document. You would actually be able to see these erasures when the document is photographed on infrared material, since traces of carbon or other material embedded in the fibers of the document's paper, which were caused by the erasures, would become visible.

Most important to the reproduction technician is the fact that infrared film permits us to bring back a visual loss in the image areas of drawings, photographs, artwork, and other types of originals (Figs. 6.6 and 6.7). For example, if you are confronted with a situation of reproducing an original tracing, blueprint, or diazo print which is extremely old and has been handled to a point where all or part of the image is illegible, infrared photography would be an invaluable process for reproducing the document *legibly*.

Again, remember: if infrared photography does not prove successful, you should try one of the ultraviolet techniques explained in Sec. 6.2 before discarding the job as hopeless.

The following is a step-by-step guide to the employment of infrared photography.

Copy Setup. In copying with infrared materials, the same techniques are employed as in copying with conventional continuous-tone materials (Chap. 5). However, there is one important difference: you must use a filter to hold back all light in the blue region of the spectrum, while permitting transmission of infrared rays to the film.

The filter may be placed over the camera's lens or over the light source, although it is usually more convenient to place it over the lens. The most suitable filters are types such as the Wratten Series 70, 87, 88, and 89. Of these, 87 provides the most satisfactory results in reproduction, since it allows the greatest transmission of infrared light.

Before exposing infrared materials, make sure film holders and camera bellows are opaque enough to hold back infrared rays. This point is not as elementary as it may seem, since older-type holders and bellows, which may seem opaque to the eye, may permit infrared light to pass through them. If infrared light penetrates the holder or bellows, it will fog the film.

Test the bellows when making exposure test shots. If the film fogs, have a camera repairman correct the leak, if possible.

If you use Eastman Kodak or Graflex cut-film holders, it is easy to determine whether they are suitable for infrared use. The slides of these holders have very fine embossed dots on top of them, which indicate they are safe for infrared use. These are the same embossed dots a photographer uses to keep track of the film he has exposed.

Focusing the Camera. The longer wavelengths of infrared rays may pose a problem in focusing, since they could cause a color defect in the lens. To overcome this problem, tests to determine the exact point of focus should be conducted.

The manufacturer of your lens will usually supply information, on request, concerning compensation factors to use when working with infrared materials. However, a rule of thumb you can employ to determine this factor is as follows:

The color defect (or chromatic abberation) of a lens when using infrared materials can be corrected by extending (bringing forward) the lens board by $\frac{1}{200}$ the focal length of the lens.

This is done *after* the image has been brought into focus on the camera's ground glass. Remember, however, that no matter how sharp the image on the ground glass, the final reproduction will not be clear and sharp unless compensation is made.

Suppose, for example, you are working with a camera equipped with a lens that has a 10-in. focal length. Bring the subject into perfect focus on the ground glass. Now, compensate for the use of infrared materials by employing the rule of thumb. Move the lens forward $\frac{1}{200}$ of the focal length of the lens (10 in.), or approximately $\frac{1}{16}$ in.

Illumination. Graphic-arts reproduction cameras employing carbon-arc lamps as illumination sources are excellent for reproducing with infrared materials. Arc lamps emit a relatively large amount of infrared rays.

Other lamps suitable for reproduction by infrared are specially designed infrared bulbs, which are red in color, or tungsten-filament lamps, which radiate large amounts of the necessary infrared rays.

Exposure. Exposure must be accurate. Unlike conventional reproduction materials, infrared films leave little room for error in overexposure and underexposure. Accurate exposure is vital, since you are trying to bring forth maximum detail which, many times, will not be visible to you until the negatives are developed.

You have already made focus and exposure tests at different scales of reduction and enlargement with your equipment (see Chaps. 1 and 5). You must now find the correct exposure when using infrared materials, since compensation must be made in exposure for the filter you are using. By using the exposure results of the tests made with conventional continuous-tone material (see Chap. 5) as a constant and testing with infrared materials to determine the difference in exposure between conventional and infrared materials, you can establish a filter factor. To determine this factor, divide the constant "conventional exposure" by the "infrared exposure."

For example, suppose with a conventional film you established a normal exposure to be 5 sec at a given lens opening. You test the infrared film and find that normal exposure at the same degree of reduction or enlargement as for the conventional film test is 10 sec. (Test procedure is the same as explained in Chap. 5.) Your exposure (filter) factor, then, is 2, and you use this factor to obtain accurate exposure with infrared films at any degree of reduction or enlargement.

Development. Infrared films are available in cut-film and 35-mm sizes. The leading manufacturer of these materials is Eastman Kodak, and their developers are the best suited for infrared photography.

Optimum development results with infrared films are obtained by using D:11 or D:19 developers. D:76 and DK:50 developers can also be employed, although they provide less than perfect results.

Infrared materials are developed in the same manner as conventional photographic films, with one important additional procedure. The antihalation backing of the film must be rubbed off *during* development, as specified by the manufacturer for the particular type of film being used. Usually, this is done by rubbing the back of the film with a moist cotton pad or a soft viscose sponge.

The safelight used when developing infrared films is extremely important and must be selected with care. Safelights suitable for panchromatic and orthochromatic films are *not* suitable for infrared.

Use a Wratten Series 7 filter in a lighttight safelight lamp with a 15-watt bulb at a distance of not less than 4 ft from the film. *This is the only type of safelight that is safe for infrared films.* It is absolutely *unsafe* for orthochromatic and panchromatic materials.

6.2 Reproduction by Ultraviolet Photography

Ultraviolet photography is the second "last resort" reproduction method to use when copying continuous-tone and line material which possess illegible image areas. Ultraviolet and infrared are at the opposite ends of the color spectrum (Fig 6.1). As mentioned before, the infrared region is from 700 to 1,300 millimicrons. Ultraviolet, on the other hand, starts at about 400 millimicrons and extends downward almost to the bottom of the spectral scale. Thus, ultraviolet, like infrared, is invisible to the human eye, which sees at a spectral response of from 400 to 700 millimicrons.

Ultraviolet light is always present when photographing, but it cannot be picked up by ordinary lenses and films in its pure state at the lower end of the spectral scale, since conventional lenses transmit light in the 300- to 400-millimicron range, and the gelatins used in the emulsions of films are almost opaque to rays having a spectral response of below 300 millimicrons.

Ultraviolet photography is extremely useful in the examination of and search for altered documents, engravings, and forms of art, such as paintings. Inks and fingerprints on multicolored surfaces that cannot be seen by eye, for example, are easily detected by ultraviolet (Figs. 6.8 and 6.9). Ultraviolet photography is also used in the lithographic reproduction process where highly color-corrected reproductions are desired.

When employing ultraviolet photography, little additional operating expense is required. Most times, you employ a standard camera, conventional films, usual darkroom facilities and, in most cases, the same lenses. The major difference in equipment is, possibly, the use of different light sources and filters.

Light sources for ultraviolet photography include carbon-arc lamps, which emit practically equal amounts of infrared and ultraviolet light, mercury-

FIG. 6.8 Ultraviolet light reveals altered registration. The photograph on the right is a reproduction of what seems to be a perfectly legal automobile registration. Photographed under ultraviolet light, however, the reproduction (*left*) shows the alteration that was done to the registration. (*Courtesy of the Federal Bureau of Investigation, U.S. Department of Justice.*)

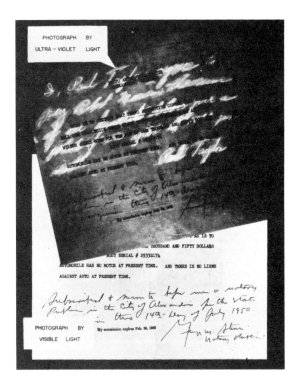

FIG. 6.9 Fraudulent notarization of bill of sale. Photographed under ultraviolet light, writing that reveals a fraudulent transaction was discovered. (*Courtesy of the Federal Bureau of Investigation, U.S. Department of Justice.*)

vapor lamps (either quartz or glass), and blacklight fluorescent lamps. In addition, many companies make various types of lighting equipment specifically designed to give ultraviolet radiation. If you wish, you could contact them to learn what is available to meet a specific problem. Leading companies in this field include UltraViolet Products, Inc., of San Gabriel, California; Lamp Division of General Electric Company, Nela Park, Cleveland, Ohio; and Hanovia Chemical and Manufacturing Company, Newark, New Jersey.

Filters for ultraviolet photography are employed only when the light source is not specifically an ultraviolet one. When using carbon-arc or mercury-vapor lamps, for example, a Wratten Series 18A filter should be placed over the camera's lens or over the light source itself. This filter passes narrow bands of light, the wavelengths of which are in or near the ultraviolet range (approximately 300 to 400 millimicrons). If fluorescent lights are used in ultraviolet reproduction, a Wratten Series 2A or 2B filter should be used. Either brings light near enough to the ultraviolet range so reproduction can be accomplished.

Exposure when using filters has to be determined by trial and error, because of the presence of unknown factors, such as the intensity of the light source and the degree to which the filter over the lens or light source will filter out wavelengths other than ultraviolet. The best film to employ with a light source not primarily ultraviolet is a commercial color-blind type (see Chap. 5) because in ultraviolet photography there is no need for a film to be sensitive to green or red.

As with infrared photography, ultraviolet photography usually requires a focus other than the clean, sharp one appearing on the ground glass. If your first exposures are not sharp using accurate focusing on the ground glass, you should compensate by moving the lens closer to the film (that is, move the lens board toward the film plane) ¹⁄₆₄ in. for every 4 in. of the lens focal length. When using a lens with a 3-in. or shorter focal length, no compensation is required.

Suppose, for example, you use a 4-in. focal length lens. First, focus the subject clearly on the ground glass. Then move the lens ¹⁄₆₄ in. closer to the film. If an 8-in. focal length lens is used, move the lens ¹⁄₃₂ in.

You can employ any camera and lens in ultraviolet reproduction. However, special lenses of quartz are manufactured specifically for this type of work. They are particularly applicable for work which requires transmission of wavelengths shorter than 350 millimicrons. In this sense, then, they are needed for jobs which present unusual and difficult conditions.

A quartz lens is a one-piece lens, while a regular lens is composed of two or more pieces of optical glass cemented together. The latter, if used for ultraviolet reproduction in the lower part of the color spectrum, may fluoresce and not transmit the ultraviolet rays. Quartz, on the other hand, which transmits more ultraviolet light than optical glass, presents no such difficulty.

There are two standard methods of reproduction by ultraviolet photography: reflected light method and fluorescent light method.

The *reflected light method*, which is the one used in "normal" ultraviolet reproduction, is accomplished in a manner similar to conventional reproduction. The subject matter is illuminated in a standard fashion, and exposure is made from light reflected back toward the film from the original being copied. The one difference is that the camera's lens or the light source is covered by a filter to hold back the visible rays of the spectrum and permit transmission of ultraviolet rays, as previously discussed; or, if you employ special ultraviolet sources, no filter is required [Fig. 6.10(1)].

Success of the *fluorescent light method* depends on the *fluorescence* of the subject. Oddly enough, with this method you wish to filter out all ultraviolet light and transmit only the *fluorescence*, which is a wave of light longer in length than ultraviolet light [Fig. 6.10(2)].

FIG. 6.10   Ultraviolet reproduction setups. Part 1 (*left*) is a typical ultraviolet reproduction setup. Components are (a) mercury-vapor lamps or other light sources; (b) ultraviolet transmission filter; (c) ultraviolet absorbing filter used in the fluorescent light method only; (d) the copy easel. Part 2 (*right*) is an alternate method for the fluorescent light technique using only one light. This setup is effective for reproduction of more contrast.

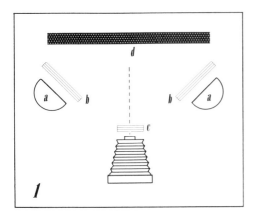

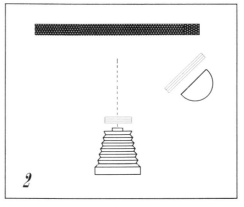

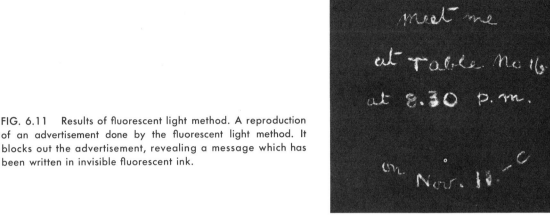

FIG. 6.11 Results of fluorescent light method. A reproduction of an advertisement done by the fluorescent light method. It blocks out the advertisement, revealing a message which has been written in invisible fluorescent ink.

Subjects set up for copying in a darkroom and illuminated by specially designed ultraviolet light sources radiate this longer wavelength. It is only this long wave that is recorded on the film. The rest of the image, including the ultraviolet light, is dropped out. The filters to use with the fluorescent light method are a Wratten 2A or 2B, which do not transmit ultraviolet light but do transmit all the fluorescent light from the subject.

The fluorescent light method is used primarily for detecting altered documents, when you wish to change the reflecting power of ink, papers, paint pigments, etc. (Fig. 6.11).

Paper, for example, is composed mainly of cellulose, which "fluoresces" strongly in ultraviolet light. If a sheet of paper has been inked and the ink has been erased or chemically removed, those parts of the paper where the ink has been tampered with will have a lower fluorescing power than the rest of the material. By use of the fluorescent light method, you can make a photo reproduction of the original inked writing, although it is invisible to the eye.

Development of ultraviolet reproductions are accomplished the same as when developing conventional films, for you *are* developing conventional films. Develop type B panchromatic emulsions in a developer such as DK:50. If employing process panchromatic or process orthochromatic film, use a high-contrast developer, such as D:11. Dark green safelights can be employed during development.

6.3 Reproduction of Radiographs

The radiograph (X ray), which has been employed successfully in medical practices for so long, has now been adopted by industry. Special X-ray equipment, designed specifically for industrial uses, is being employed to detect cracks and breaks in metal, for photographing the insides of metal castings, and for many other specialized industrial applications.

With the extension in use of X-ray techniques to industry, the reproduction technician finds that his work load is on the rise. Often, requests for reproductions and duplicates of radiographs are needed for projection slides, display transparencies, and distribution, as well as paper prints of X rays for illustrative and publication purposes.

Radiographs are continuous tone in nature and are reproduced on continuous-tone materials. The originals to be reproduced are, of course, different than the subject matter from which they were made, since they are in negative form and must be reproduced as negatives, but the procedures for making the reproductions differ only slightly from conventional processes.

The principles outlined here will provide you with all necessary information for instituting a radiograph reproduction capability in your laboratory. One word of caution, however, is advisable at the outset: reproduction of radiograph originals must be accurate, or a valid interpretation and analysis from the duplicate will not be possible. The criterion for accuracy is the tonal range of the reproduction, and not necessarily detail of the reproduced image which may not be as sharp and as well defined as other types of originals.

The method used by most reproduction technicians for copying radiographs is called the *intermediate negative method.* Another method is solarization, but this is employed primarily by X-ray technicians and will not be explained here, since it has limited application to general reproduction departments.

The intermediate negative method is the best all-round procedure for reproducing radiographs. It provides the most accurate way of accomplishing this specialized reproduction process. To employ this method, the equipment needed for continuous-tone reproduction (Chap. 5) is required. Primarily, this equipment includes a graphic-arts reproduction camera with a transparency opening or a view or press-type camera coupled with a transparency illuminator or a light box.

The intermediate negative method is accomplished as follows:

1. Place the original radiograph to be copied on a transparency illuminator or a light box, or in the opening of a graphic-arts reproduction camera's transparency window. Mask the entire area with black paper, so that light passes only through the image area, eliminating all glare.

Handle the original radiograph with extreme care. Dust specks, particles of lint, and other foreign matter should be cleaned off carefully with a fine camel's-hair brush or a soft photo chamois before mounting it on the illuminator for reproduction. Remove fingermarks and other smudges with a piece of absorbent cotton moistened in a film cleaner. Marks or foreign matter on the original radiograph, if not removed, will be accentuated on the reproduction, impairing the effectiveness of the reproduced tonal range.

2. Turn on the lights of the transparency illuminator or light box. If you are using the transparency opening of a reproduction camera, move your lights so illumination passes through the radiograph. It is this transmitted light by which you make the reproduction.

Check your illumination source for even lighting over the original's entire surface. Infrequently, illuminators such as light boxes and transparency viewers provide sufficient light for viewing, but light of an uneven nature for photographing. If this is the case with your setup, try to obtain a more evenly lighted viewer or use a reproduction camera.

3. In focusing, be aware of the fact that it is often difficult to obtain an accurate focus from a radiograph original, since the image is rarely sharp. Focusing by use of a copy camera with automatic scale settings, however,

presents no problem. However, if a view or press-type camera is employed, a focusing aid will be needed.

Place a small sheet of clear acetate containing some ruled lines over the radiograph original. By focusing on the lines, you will obtain an exact and accurate focus, which is as sharp as possible.

4. Use an orthochromatic film of medium speed and contrast to achieve the best results. Commercial-type films, Kodak gravure copy film, and Du Pont XF ortho film (type 421), are suitable.

5. Correct exposure should be determined by test, since there is no accurate way of taking a meter reading when employing this type of reproduction method. The method of accomplishing these tests outlined in Sec. 5.4, applies to the reproduction of radiographs as well.

6. Develop test negatives in solutions such as Kodak D:76 (undiluted), Du Pont 6D, Ansco Isodol (diluted 1:1), or Kodak DK:50 (diluted 1:1) for the recommended time. In most cases, you will find this recommended development time is not the one that produces negatives of proper contrast, and you may have to underdevelop the negatives to obtain normal contrast. After determining both correct basic exposure and development time, it may be useful to compose an accurate time-temperature chart for developing radiograph reproductions.

A properly exposed and developed radiograph reproduction is one that has a slightly higher density, but a lower contrast, than a normal continuous-tone negative. Overexposure obscures detail on the reproduction in low-density areas copied from the original. If the reproduction is underexposed, detail in high-density areas will not be recorded.

7. You can now take the intermediate negative and print it in a number of ways and on a number of materials to produce the type of duplicate desired. The negative can be projected in a photographic enlarger, or it can be contact-printed on conventional contact papers.

When making enlargements or contact prints, be wary of overall contrast, and try to approximate the tonal qualities of the original. If the negative has been exposed and developed properly, all prints will be produced with excellent results on papers of normal contrast.

In making display transparencies from the intermediate negative, project or contact print the negative onto materials such as Kodak fine-grain positive film, Du Pont Adlux, or any commercial-type film. Develop Kodak fine-grain positive material in an MQ-type developer diluted 1 part of developer to 3 parts of water. This material can be handled safely under red light. Develop Du Pont Adlux in Du Pont 53D, 55D, or any all-purpose MQ-type developer diluted 1 part of developer to 2 parts of water. Adlux can be handled safely under a Du Pont S-55X safelight filter.

part III | TECHNIQUES OF ENGINEERING REPRODUCTION

7 | THE DIAZO AND BLUEPRINT PROCESSES

UNLIKE OTHER REPRODUCTION METHODS discussed thus far in this book, the diazo and blueprint processes utilize special machines, and not conventional photographic and lithographic equipment, to make the actual reproduction.

A diazo or blueprint is a reproduction of an original or a reproduction of an intermediate which the reproduction technician or photographer is responsible for making. (See Chap. 8 for a discussion of the preparation of duplicate originals and intermediates.) By being familiar with the diazo and blueprint processes, the reproduction technician is in a better position to understand the characteristics of the specialized machines used for making the reproduction, thus enabling him to prepare the type of duplicate originals and intermediates that will provide final diazos and blueprints of the highest quality.

In many companies, the diazo and blueprint function is the direct responsibility of the reproduction department. It is not unusual for a reproduction technician to be called upon to operate a diazo or blueprint machine in addition to his other reproduction duties. In those companies where the diazo or blueprint function is separated organizationally from the reproduction department, close coordination and cooperation between the two must be maintained.

Diazo and, to a lesser extent, blueprint are versatile processes that can be used for many tasks besides simple plan copying. By being familiar with their principles and operational characteristics, the reproduction technician can use them to good advantage by combining them with other reproduction methods to make neater, more powerful, and more functional presentations.

The primary use of diazo and blueprint reproduction has been in the area of copying engineering drawings. In this alone, millions of pounds of diazos and blueprints are produced each year. Those in the diazo and blueprint industries often brag that if a superjet aircraft were loaded with all the diazos and blueprints used in its construction, it would never get off the ground; or, if the compartment of an aircraft carrier were loaded with its respective diazos and blueprints, it would sink at the dock.

Diazo and blueprint are grouped together in one chapter since they both provide a same-size duplicate (facsimile) of an original by means of a semiautomatic or completely automatic machine. Here, however, the similarity ceases.

Although the term *blueprint* is often erroneously applied to diazo reproduction, a diazo is *not* a blueprint, and a blueprint is *not* a diazo. The two methods are different in principle, in operation, in end result, and in application, and to use one term that applies to both is a serious mistake which could lead to the wrong end product. To eliminate any confusion, therefore, each is discussed in this chapter under a separate section heading.

There are many different types of diazo and blueprint machines. It would be impractical, not to mention impossible, to discuss all makes. Thus, we confine our discussion primarily to the principles of each process and leave to the manufacturer of each machine the responsibility of instructing his customers through manuals as to the methods of employing his machine.

Significant advances in diazo have caused blueprinting to assume a secondary position. Diazo offers many advantages not inherent to the blueprint process. Some of these are its greater economy and efficiency; the more desirable positive-to-positive reproduction; simpler machines to operate and less room needed for their installation (there are diazo machines available that are small enough for desk-top use); and the greater versatility offered through a greater variety of sensitive materials. In addition, diazo machines are available in less expensive units, while most blueprinting equipment is costly and expensive to install.

The diazo process is a more flexible process than blueprint. With it, one can produce color images on acetate for use as overlays on artwork and for large format slides in overhead projectors. (The use of diazo for producing smaller-size slides, such as 35 mm and 2¼ by 2¼ in., is not recommended since there is a noticeable loss of detail when reproducing these smaller presentations in a diazo machine.) Certain diazo materials permit almost identical looking reproductions of continuous-tone photographs, while others are available on dimensionally stable bases that offer extreme stability of the image. There are even offset plates on which a negative can be exposed in the diazo machine for small presses.

Most blueprint facilities, recognizing the advantages of diazo, have added a diazo capability. This, however, does not mean that blueprint has become extinct. Although it has lost much of its importance, there are still certain professions and industries, not to mention governments, that require reproduction of drawings by blueprint. For this reason alone, we cannot disregard blueprint, and it is discussed in Sec. 7.2.

7.1 The Diazo Process

Basic Principles. The diazo process (also referred to as whiteprinting and diazotype) produces a same-size reproduction of an original in a direct positive manner; that is, it produces a positive reproduction of a positive original, and a negative reproduction of a negative original.

The term *diazo* is derived from the basic compounds that make up the sensitive material's emulsion, which are diazonium salts. When these salts are exposed to the rays of an ultraviolet light, they decompose (bleach) and are unable to couple with a developer, leaving an area that is white (bleached).

In operation (Fig. 7.1), an original that has a translucent or transparent base is fed into the diazo machine in contact with a sheet of sensitized diazo material, with the original on top of the diazo material and its image facing up. The two pass around a glass cylinder, which houses an ultraviolet light source, usually a mercury-vapor quartz lamp. The opaque image areas of the original prevent the light rays from striking the diazo material, while the clear background areas of the original permit the light rays to penetrate through to the diazo material.

Those portions of the sensitive material struck by light are deactivated, that is, the diazonium salts are decomposed. The diazonium salts remain unaffected in those areas not exposed to the rays of the light.

When the diazo material is passed into the developing chamber of the machine, those areas not affected by the light—the image areas—assume a dye color through a process known as *color coupling*. The alkaline developing medium couples with the unexposed diazonium salts of the diazo material to form a dye color, such as blueline, blackline, or sepia, depending upon the type of material being used.

Those areas of the sensitive material that were decomposed by light during exposure are not affected at all by the developing medium and remain white.

The first diazo reaction was produced in 1858, with the first diazo direct

FIG. 7.1 The diazo machine. Note its fully automatic features. (*Courtesy of the Ozalid Division, General Aniline and Film Corporation.*)

positive being made in 1890. It was not until World War II, however, that the full potential of the process was realized. Since then, many businesses and industries, particularly engineering, drafting, and architectural firms, have come to rely almost exclusively on diazo reproduction. As has already been mentioned, technical advances in diazo machines and products have evolved to a point where the process is considered more economical, efficient, and practical to employ than blueprinting.

The Original. Just as with any reproduction process, diazo demands that the original for reproduction possess certain characteristics. These take the form of the material on which the original is made and the composition of the image itself.

The *material* on which the original is made for reproduction by diazo must be translucent or transparent in nature. Translucent and transparent materials permit light to shine through the background areas of the original to the comparable areas of the diazo sheet beneath. The only areas of the original that should block light from reaching the diazo material are opaque image areas.

Unless sufficient light strikes the nonimage areas of the diazo sheet, no image will form. When struck by light, these areas decompose and provide a white background to offset the nondecomposed (or image) areas of the diazo material. Unless the original is composed on suitable material, permitting light to shine through it to the diazo material, the diazo developer will turn the entire diazo sheet dark.

In cases where originals are not on a translucent or transparent base, it becomes necessary to make a duplicate original, commonly called an *intermediate* (or *second generation reproducible*), on this type of base. This operation is usually accomplished photographically as explained in Chap. 8. However, an intermediate can also be made nonphotographically by use of the diazo machine itself, and this is frequently done, even though the original is on a suitable base, to protect the original from excessive handling.

Called a *sepia intermediate*, the machine-produced duplicate original is made on a diazo sepia sheet, which has a translucent base. As the name implies, the image produced is brownish-sepia in tone. If made properly, it can be used time and again for making diazo reproductions of good quality.

The use of the sepia intermediate offers two distinct advantages. Although the original must be produced on a translucent or transparent base to make a sepia intermediate, you may not want to use that original for constant reproduction through the diazo machine. The more times an original is run through the machine, the greater the chances of its becoming creased, wrinkled, or damaged by the heat generated in the machine. Thus, to protect the original, which may have to be sent to a customer or kept on file, you run it through the machine only once to make a sepia intermediate. Then, you use the sepia intermediate to make as many reproductions as required.

A second advantage of the sepia intermediate lies in its flexibility for changes. Suppose, for example, you have a job where a number of drawing alterations are necessary. These drawing alterations are all modifications of a basic drawing, differing from that drawing only in titles, callouts, or minor changes in the illustration itself.

To save drawing time, the basic drawing can be reproduced a number of times on sepia intermediate material. Then, revisions can be made to the sepia intermediates through use of a chemical eradicator and black ink or drawing pencil. The revised drawings can then be run through the diazo machine to make as many reproductions as are necessary.

Sepia intermediates can be made in one of two ways. First, they can be made in the same manner as any diazo reproduction, that is, with a right-reading image positioned the same as the image of the original. To produce a sepia intermediate in this manner, the original is put on top of the sepia material with the image facing up. It is then run through the machine. This method, however, does not produce as sharp an intermediate image as the reverse-reading way of producing a sepia intermediate.

To make a reverse-reading intermediate, position the original on top of the sepia intermediate, but with the image facing *down*. The result is a print with a backward (reverse-reading) image. To use this type of sepia intermediate, place it on top of a sheet of diazo material with the image facing down (that is, emulsion to emulsion). This produces the best possible contact between intermediate and sensitive sheet and an image of the highest resolution.

The *composition* of the original's image must be dark enough to prevent light from reaching the diazo paper. It is therefore advisable to prepare drawings for diazo reproduction with black drawing ink or dark black pencil. However, most other colors used in drawing an image will be suitable, with the exception of shades of blue.

Although blue may appear opaque to the eye, it transmits ultraviolet light and has little or no printing value when used in diazo. For this reason, many types of drafting materials have light blue grids or preprinted guidelines on them. These lines are used as guides in drafting with the assurance that they will be completely lost when the original is reproduced by diazo. By comparison, relatively light colors used on an original, such as yellow, absorb ultraviolet light and, therefore, will reproduce properly on diazo material.

You may run into situations where the background of an original, although made on a semitranslucent base, is too opaque to transmit sufficient quantities of ultraviolet light to the diazo material beneath, resulting in exposures that leave much to be desired. In these cases, you could probably transparentize the original by applying a thin coating of transparentizing solution to the back of the original. This solution is merely an oil-base liquid which is applied sparingly with a rag or swab of cotton. It makes the original translucent, much the same as cooking shortening or oil poured on a sheet of paper makes that paper translucent.

The transparentizing effect is permanent, but will not stain the original although its color will be changed slightly. Transparentizing can be used effectively on lightweight papers of about 20- or 30-lb bond. It is not effective on opaque stock or cardboard. To determine what the effects are on a particular original, test the material in question by applying transparentizing solution to a piece of scrap or a corner of the original itself.

Originals printed on both sides should never be transparentized, since the image on the reverse side will print through when the original is reproduced by diazo. Typing papers can be run through the diazo machine without transparentizing if the stock is very thin, such as onionskin. More opaque bond typing paper, however, will produce poor-quality reproductions if not transparentized, since their light transmission qualities are low.

If typewritten material is submitted for diazo reproduction, the reproduction technician should recommend to the originator that the original copy be carbon-backed with orange-colored carbon. This material provides dense impressions on the back of the original, which results in good, sharp, clear reproductions that hold back ultraviolet light.

The Two Diazo Processes. There are actually two types of diazo processes: the semidry developing process (moist process) and the dry developing process (dry process). Essentially, both processes are the same in their mechanics and the results achieved. The differences are in the method of development and the materials used.

Materials for the moist diazo process are coated with a light-sensitive diazo emulsion only, while materials for the dry diazo process are coated with a combination light-sensitive emulsion and coupling agent.

In the *moist process,* the coupling agent necessary to form the reproduced image is contained in the diazo machine in the form of a liquid developing solution, which is applied to the surface of the sensitive material when it is passed into the developing cylinder after exposure. The solution contains alkaline salts that couple with the diazonium salts of the sensitive material to form a permanent, visible colored image. To change the color of the image, one must change the developing solution in the machine.

In the *dry process,* the sensitive material is developed by passing it through an alkaline ammonia vapor, which combines with the coupler in the sensitive material to form the permanent, visible colored image. One changes the color of the image by using different types of sensitive materials, thus making the dry process more flexible than the moist. To produce images of different color—blueline, blackline, and sepia tone—all an operator has to do is simply change the material he is using.

The Diazo Machine. Machines for the dry and moist procedures are different in the construction of their development chambers.

Dry-process machines are composed essentially of two chambers: exposure and development. After exposure, the sheet of diazo paper is fed through a series of rollers in the development chamber where it is subjected to the fumes of the developer. The development chamber is basically a large, stationary tank, with perforations on one side through which vapors are emitted.

The vapors that do the actual developing are produced by liquid aqua ammonia dripping on an electrically heated tray outside the developing chamber. The rate of ammonia evaporation and, thus, the density of the vapor depend on the amount of liquid ammonia permitted to drip on the tray and the temperature of the electric heater.

Most dry process units have controls that enable the operator to regulate the amount of aqua ammonia dripping on the heater tray and the heat of the tray itself. In addition, these units also possess a control that permits the operator to adjust feeding speed and another control that automatically adjusts the rate of developing vapor seeping into the development chamber.

Standard diazo machines can be converted to automatic ammonia-injection systems that use compressed gases rather than liquids. Development results are the same, however.

Machines for the *moist diazo process* employ rollers through which the exposed material is passed. These rollers apply a thin film of liquid developer evenly over the material's entire surface, resulting in conversion of the unexposed diazo coating to an azo dye image that is identical in appearance and permanence to the dye image produced by dry-type machines.

Diazo dyes are soluble in water and other liquids. Since the image will wash away if it comes in contact with any liquid, it is advisable to keep diazo prints away from water and other liquids.

In summary, the following is a step-by-step account of what happens in both the moist and dry diazo processes:

1. The original is placed in contact with the sensitive material, with the original on top and its right-reading side facing up.

2. The two sheets are fed into the machine, where they are kept in close contact with each other, usually by a suction system that holds both sheets together on a revolving rubber or canvas belt (Fig. 7.2).

3. The original and diazo material pass around a glass exposure cylinder, where a high-intensity ultraviolet light makes the exposure.

4. The original and exposed material are then returned to the operator. (Many machines return only the original to the operator and automatically pass the diazo sheet into the development area.)

5. The operator feeds the exposed material into the development chamber, where the unexposed compounds are developed to form the dye image.

6. The final reproduction is returned to the operator.

The ease with which an operator can perform the above steps is based upon the complexity of the machine employed and the degree to which it is automated. Most larger units possess complete automatic features that are controlled by the operator through various external adjustments. Most portable units, however, are manually operated and require extra procedures for development, such as the so-called "pickle-jar" method.

The "pickle-jar" method is employed when the diazo machine does not

FIG. 7.2 Diazo feeding belts. These are the two belts the operator sees when operating his diazo machine. The one on the bottom pulls the original and sensitive diazo paper into the exposure chamber. In most machines, the original is returned to the operator on the top belt, while the exposed material continues around the belt into the development chamber.

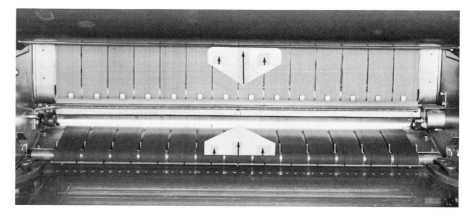

possess a separate development chamber. A glass or plastic jar or tube, separate from the machine itself, is used. This jar or tube contains a sponge on its bottom, which is saturated with ammonia. When the exposed print is placed in the jar or tube, the top is closed, and the ammonia fumes accomplish development.

Most diazo machines possess speed controls that regulate the feet per minute at which the machine can print, that is, the total time necessary for the copy to be exposed. The faster the printing speed, the quicker the exposure, and the darker the image and background will be. In other words, when the sensitive material is rapidly exposed to the light source, the diazo coating in the background areas has little time to decompose. Thus, the background may take on an overall dark appearance, which may not always be desirable. An ideal diazo reproduction is one with a clean, well-developed image area on a clear, white background.

Conversely, a slow (longer) exposure produces a lighter image area and background, since more bleaching occurs to the entire surface of the sensitive material. If the exposure is too slow, the image will be too light.

In the diazo operation, the final appearance of the reproduction is affected by exposure, and not by development. The latter is, more or less, a constant. The development controls on the machine permit an operator to guarantee that there is sufficient vapor in the development chamber to develop the image or, in the moist process, that there are sufficient quantities of liquid ammonia on the rollers. They do not permit an operator to make an image area lighter or darker.

Situations may arise where a compromise in the quality of the reproduction is necessary to produce a legible image. In the case of originals drawn in pencil, for example, there may be occasions where the line density of the image is so low that light would penetrate the image in exposure. In cases such as this, a reproduction will have to be made at a fast printing speed, so that the image is not broken up (too light) on the final reproduction. Thus, the image may appear darker than desirable, but it may be the only way to preserve detail.

A rule to remember when printing difficult originals is this: Adjust printing speed to get maximum legibility with a minimum of background color.

More than anything else, it takes experience to make an efficient diazo-machine operator. Experience will teach an operator how to handle the various types of originals in his machine and will provide a basis for his handling of various types of originals.

A valuable aid for an operator is his own testing procedure. He should run a few sample tests using originals or intermediates that are typical of his work load. Then, he can draw up an exposure chart that lists the basic exposures for ink drawings, pencil drawings, sepia, typewritten sheets, and so forth. By varying these exposures slightly, he could probably get good reproductions of any type of original.

Diazo Materials. There are many different types of materials available for diazo printing. These differ as to image color, such as blueline, blackline, and sepia, and as to the stock on which the coating is placed, such as linen, Mylar, cardstock, and clear acetate.

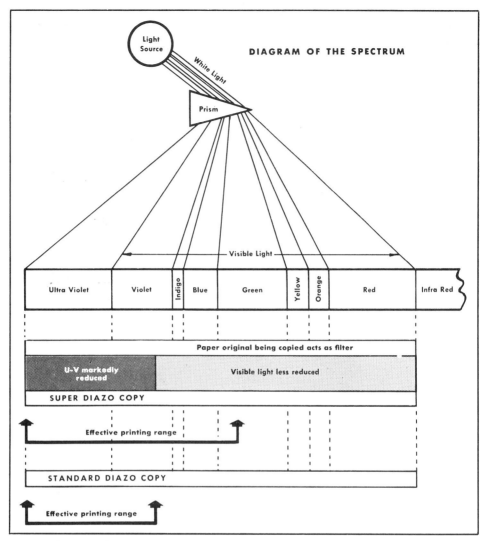

FIG. 7.3 Diazo spectrum. The spectral response of standard diazo materials is shown on the bottom, while the color spectrum is seen on top. The widened range of new, more highly sensitized materials in their relation to the spectrum and to standard diazo materials is shown in the middle. (*Courtesy of Frederick Post Company.*)

As was mentioned before, these materials are sensitive to high-intensity ultraviolet light, but can be handled for a reasonable period of time in room light, even if this room light is fluorescent, which is high in ultraviolet radiation (Fig. 7.3).

However, it is advisable that you do not expose diazo material to room light unnecessarily. If in room light for too long a period, the diazo coating will begin to decompose.

To be absolutely safe, many installations use yellow light as their illumination source in a work area. Yellow light is practically devoid of ultraviolet radiation, which is filtered out by the yellow glass of the bulb or tube.

Unexposed diazo materials are adversely affected by humidity. For this reason, it is recommended that materials be stored in a controlled humidity atmosphere, such as an air-conditioned area. Naturally, the unexposed

FIG. 7.4   Color foil used in flip charts. Diazo color foils are often used as overlays for flip charts and other training aids. The desired overlay material is reproduced on the color foil, and it is then laid over the drawing. (*Courtesy of Lockheed Electronics Company.*)

material should be kept away from an area when ammonia is being used, especially the area in which the diazo machines are operating. Ammonia fumes hitting the material will cause premature development.

It is a good idea to store diazo paper in its original package or in containers that are fairly lighttight until it is needed and after it is used. This procedure guards the emulsion against possible decomposition by natural room light.

As we have indicated throughout this section, diazo offers a great deal of flexibility and versatility. To illustrate, the following is just a partial list of the various diazo materials which are available:

Color Foils. These are transparent acetates that produce various colored images. They are widely used for color overlays to be employed in overhead projectors and for overlays on training aids and charts (Figs. 7.4 and 7.5).

Continuous-tone Materials. When placed in contact with conventional photographic negatives and run through the diazo machine, these produce prints that look like photographic prints.

Microfilm Duplicating Materials. These are available in 35-mm and 16-mm strips, and are designed for low-cost duplicates of microfilms.

Diazo-coated Cardstock. Available in various colors, these are used as page separators and for other requirements where a rigid stock for color coding is needed.

Stable-base Materials. These are excellent materials for holding exact size, where a great deal of permanence and stability is needed.

Cloth-base Materials. Available on a heavy canvas base, these materials are very often used for flip charts.

Offset Plates. These are exposed in the diazo machine instead of an arc printer and are developed with special developing fluid. They are then put on an offset duplicator for reproduction of many copies.

7.2   The Blueprint Process

Blueprint is a negative-positive process that produces a same-size reproduction of an original. It provides a right-reading positive image from a

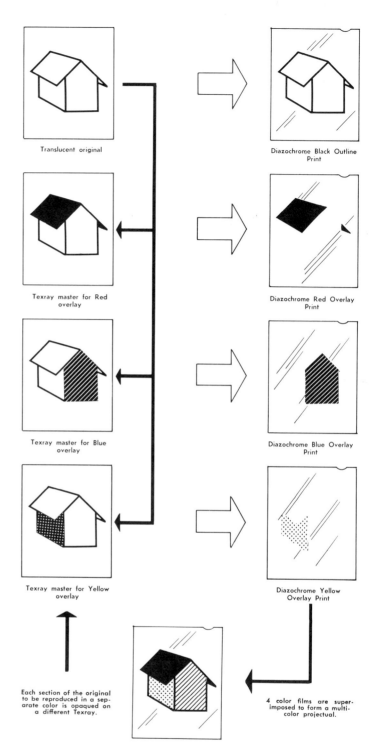

FIG. 7.5 Making color-foil slides for projection from black-and-white translucent originals. The column on the left illustrates various color overlays in place, while the column on the right indicates what each translucent original looks like. The bottom figure is the completed slide. These color acetate foils are printed in the diazo machine from a blackline translucent drawing. (*Courtesy of Technifax Corporation.*)

Translucent original

Diazochrome Black Outline Print

Texray master for Red overlay

Diazochrome Red Overlay Print

Texray master for Blue overlay

Diazochrome Blue Overlay Print

Texray master for Yellow overlay

Diazochrome Yellow Overlay Print

Each section of the original to be reproduced in a separate color is opaqued on a different Texray.

4 color films are superimposed to form a multicolor projectual.

negative original and a right-reading negative image from a positive original. The common use for blueprints is for construction and engineering drawings. These are negative reproductions of positive originals and produce a print that has a white image on a blue background. By comparison, a positive reproduction of a negative original would produce a blueprint with a blue image on a white background.

FIG. 7.6 Photograph printer. This is a special diazo printer used to make red print-cut proofs from continuous-tone negatives. The machine does not have a development chamber, since the paper used in the process does not need chemical processing. The action of the ultraviolet light in the machine on the sensitive paper produces a visible reddish color. (*Courtesy of Reproduction Engineering Corporation.*)

Blueprinting is an iron salt process that is used primarily for copying mechanical drawings. A piece of sensitive, unexposed blueprint paper is off-white in color and is coated with these iron (ferric) salts. When the exposing light strikes this paper and exposes the salts, it turns the entire sheet blue, except for those areas which block the light. These areas remain white.

To ensure that the unexposed area will remain white and not turn blue when exposed to room light during use, the exposed sheet of blueprint paper is washed in running water. The water dissolves and washes away the unexposed iron salts, which actions, in effect, fix the area much the same as a fixing bath does in making exposed areas of photographic film permanent.

Discovered by Sir John Herschel in 1842, the blueprint process has been employed for many years for copying line drawings and printed matter. It was the major copying process in use until diazo reached such a position of prominence.

Most blueprinting is performed by professional reproduction companies, which utilize large, expensive machines. This blueprinting equipment is not available in small, compact units, as are diazo machines, making it impractical for any organization other than a large reproduction facility to consider the installation of such machines. Requests for blueprinting, therefore, must usually be sent to commercial blueprinting firms.

Blueprinting papers, as diazo papers, can be handled in their unexposed, sensitized state under ordinary room light. We already know that a positive original will produce a negative reproduction. However, if it is desirable to obtain a positive reproduction (blue image on white background) from a positive original, one must first make a sepia intermediate or vandyke (brownprint). (The preparation of vandykes is discussed below.) It is this intermediate that is fed through the blueprint machine to produce a positive reproduction.

In exposing blueprint material, the original is placed face up on top of the sensitized paper and is fed into the machine's exposure chamber on a moving belt. Light from a high-intensity mercury-vapor lamp or other high-intensity ultraviolet source, such as arc lamps or fluorescent tubes, passes through the original's clear areas to the sensitive material. After exposure, the original is returned to the operator, but the exposed blueprint material continues into the washing chamber, where it is washed in water, and then into the drying chamber.

Since blueprint paper is available only in rolls, it is most economically employed when reproducing many copies of large-size drawings. It becomes costly to print individual, small-size originals because of the considerable waste factor.

You could establish a small blueprint operation in your facility, if you wished, without purchasing elaborate machinery. The first step of the process, exposure, can be accomplished by exposing original and sensitive material with an arc lamp or a bank of fluorescent lamps. The original and sensitive material should be placed in a vacuum frame for good contact.

The second step, washing, can be done in a photographic drum washer or, for that matter, in a pan. The third step, drying, can be accomplished by means of a photographic print dryer.

This makeshift operation would prove practical only for limited production of blueprints of limited-size originals. Any other type of blueprint reproduction employing this method would prove too costly.

Another popular so-called iron salt process, brownprinting (also referred to as *silver print* or *vandyke* after the British army engineer who discovered it), is similar to blueprint. It, too, is a negative-positive process, producing a negative reproduction from a positive original and a positive reproduction from a negative original. In its negative form, the reproduction consists of a white image on a dark sepia background, while in its positive form, the image is sepia in tone and the background is white.

Brownprinting is used primarily for making inexpensive negatives for use in other processes, such as blueprint or diazo. The primary way of exposing brownprint material is in a vacuum frame by means of a high-intensity ultraviolet light source. However, commercial firms that produce large quantities of blueprints may have their blueprinting equipment set up for the brownprinting process. A sheet of brownprint paper can be run through a blueprint machine in contact with a positive original to form an intermediate. This intermediate, which is in negative form, can then be rerun through the blueprint machine, or, for that matter, a diazo machine, in contact with the sensitive material for the respective machine. In the case of blueprint, the result is a positive reproduction, that is, a blue image on a white background. In the case of diazo, the result is a negative reproduction, that is, a white image on a dark background.

Exposure and development of brownprints are the same as for blueprint. Brownprint material is exposed and developed concurrently by means of a high-intensity ultraviolet light source. The fixing step, however, differs. Whereas blueprint is made permanent by washing in water, brownprint must be immersed for approximately 5 min in a special fixing solution after exposure. It is then washed and dried.

The keeping properties of brownprints are equal to those of blueprints and diazo prints, and brownprints can be used as reference prints in a shop or office without danger of being destroyed by ordinary room light. Brownprints, however, are slightly more expensive to produce than blueprints, primarily because of the additional fixing step, and they are considerably more expensive to produce than diazo prints. Brownprints are, however, usually of higher quality than the diazo sepia intermediates discussed in Sec. 7.1.

Brownprint positive reproductions of line negatives are often used by art departments in the preparation of final mechanical layouts that are to be rephotographed and made into offset or engraving plates. These reproductions provide excellent contrast for camera reproduction and possess fine detail characteristics.

Brownprint materials are available on opaque cloth or paper and on translucent base materials.

8 | PHOTOGRAPHIC METHODS OF INTERMEDIATE PRODUCTION

AN INTERMEDIATE (also referred to as a *reproducible*) is a negative or positive reproduction of an original. It has various functions, as explained below, but is primarily employed as a medium for the production of additional prints when it is not advisable or feasible to use an original drawing for this purpose.

Technically, you could say that a negative of a continuous-tone original is an intermediate, since one can make additional prints from it by enlarging or contact printing. However, in reproduction terminology, an intermediate refers to a reproduction of a line original, usually a drawing, which is used to make extra prints. Most often, but not always, these prints are produced by running the intermediate through a reproducing machine, such as diazo or blueprint.

The term *intermediate* also applies to a halftone negative that is combined in a single format with a line negative to make a single medium for obtaining combination line and halftone copies. This type of reproducible, known as a *photodrawing,* is discussed in Sec. 8.6.

When one attempts to define intermediate, it becomes necessary to categorize the functions of this medium, for it has many. These functions are as follows.

As a Duplicate to Save the Original Drawing, and for Distribution. In many industries, especially engineering, a high value is placed on original drawings, and steps are often taken to minimize the danger of damage to them by cutting down on the number of times they are handled. It is usually undesirable, for example, to use an original drawing in a machine shop or laboratory, where it can become dirty, torn, creased, or otherwise damaged. To minimize handling, therefore, an intermediate is usually made and is used as a duplicate of the original or as a medium from which copies are made for distribution.

As Second Originals That Become New Drawings. If it were not for intermediates, the many engineering changes required in process and product development and in new designs of old processes and products would necessitate constant redrafting of all or a portion of the original drawings, resulting in excessive time and money costs that are wasteful. An intermediate of an original drawing that is to be changed permits eradication of undesirable portions and additions of new data. The intermediate is then used to make as many copies of the modified original drawing as may be required, including one for filing as the modified (or new) "second original."

As a Means of Retrieving Deteriorated Drawings. Reproduction departments are often called upon to copy an original drawing (or an intermediate of that drawing), which has lost its usefulness because of a loss of line density, accumulation of smudge and dirt marks, torn or frayed areas, or similar damage. Instead of redrafting the original, reproduction of that damaged original to a usable intermediate, free of damage, is made possible through employment of certain materials. Any visible defects in the intermediate image can be touched up easily. Thus, the original can be retrieved in intermediate form at a cost considerably less than redrawing the original.

There are many ways of producing intermediates. In Chap. 7, two non-photographic methods were discussed: diazo sepia intermediates and brownprint (vandyke). In this chapter, we are concerned with the more popular photographic methods. Selection of the method to suit the specific requirements, however, could be quite complicated.

When a customer requests the preparation of an intermediate, the first task is to learn what qualities the end product must possess and, consequently, those qualities that can be forfeited. Once this is determined, the job of applying the correct intermediate production method and material to suit the customer's needs is made easier.

There are many requirements that must be considered before one produces an intermediate. The more important ones are listed below. This list can be used as a guide when questioning the customer as to his needs.

1. Must the intermediate possess exceptional *line density?* Line density of the intermediate depends to a major degree on image opacity of the original. If maximum line density is a requirement of the intermediate, the method of reproduction depends on the opacity and color of the original image. For example, an original drawn in a sepia line is best reproduced by diazo, while a less dense pencil image usually lends itself best to camera reproduction.

2. Must the intermediate be reproduced on a *transparent or translucent medium?* This is a major requirement if the intermediate is to be used for production of additional prints.

3. Should the intermediate possess maximum *line permanence of the image?* Obviously, image permanency must be sufficient to last the expected life of the intermediate. However, the extent of this life is a consideration, since it dictates the method of reproduction and, consequently, has a bearing on

cost. Intermediates produced on a photographic emulsion, which is more expensive than other means, provide reproductions of life-long permanence if processed properly. Machines, such as diazo, also provide permanent prints, but not to the same extent as photographic emulsions.

4. Should *background discoloration* of the intermediate be held to a minimum? This could be a factor if background discoloration becomes appreciable and approaches the density of the printed lines, especially if reprints are to be made from the intermediate.

5. Is *eradication* of the intermediate image essential? If a portion of the intermediate image is to be removed for application of a drafting change, the reproduction technician must select an intermediate material that provides a rapid means of eradication, with the least amount of damage to the material itself. Eradication is accomplished by one of several methods, including chemical bleaches, solvent-type removers, erasers, and scrapers.

6. Must the intermediate possess *drafting* qualities? This parallels the requirement for eradicability. Whether it be paper, cloth, or film, an intermediate material for this purpose should approach the quality of a good drafting medium.

7. Is *durability* important? Durability of the intermediate is often more important than durability of the original, since the former is usually used under more rigorous conditions. The reproduction engineer has at his disposal certain tests that he can employ to check the durability of the intermediate material. He can subject the material to water, ultraviolet light, and heat. No test is as conclusive, however, as subjecting the material to the actual conditions under which it will be used.

8. Must the intermediate have *dimensional stability?* This has become a major requirement of intermediates, especially when they are used for cartographic color-separation drafting and as nondimensional tooling drawings. Even intermediates used as engineering drawings should have a degree of stability for establishing working scales.

9. What is the *time* requirement? The production rates of modern industry often demand the shortest possible time lapse between completion of original drawings and availability of shop prints. Under these conditions, it is obvious that intermediates must be produced in a minimum of time. Various types of intermediate materials, such as those used in diazo reproduction, lend themselves to rapid processing and should be considered when time is of the essence.

10. What *equipment* is available for making intermediates? Despite other requirements, on-hand equipment (or a lack of it) often dictates the quality of the end product.

11. What are the *skills* of the technicians involved in producing intermediates? As with equipment, this could be a limiting factor. The greater the technician's experience and training, the greater will be his ability to work with the various types of intermediate materials. You will find that the materials which demand the greatest skills are also those that produce the finest intermediates.

12. *Cost* is always a factor. Very often, cost will dictate the type of intermediate-producing method employed, no matter what the requirement for quality. The reproduction engineer may often be asked to select a method that results in the least expensive intermediate and not the best.

FIG. 8.1 Setup for direct contact printing, with light source mounted in the ceiling. (*Courtesy of Lockheed Electronics Company.*)

8.1 Direct Contact Printing Method

Direct contact printing is a basic photographic method for making an intermediate. Essentially, the process involves the making of a contact print from a negative in a contact printer or vacuum frame. As long as the contact print is on a translucent or transparent material, it can be used to produce other prints by running it through a diazo or blueprint machine. Thus, the contact print serves as an intermediate.

The direct contact printing method offers one distinct advantage: the original, which may be a drawing many feet in length, can be *reduced (or enlarged)* in a copy camera onto a paper or acetate film-base material. After the negative is prepared, there are many materials to use for the intermediate. These materials for contact intermediates are available in rolls and sheet sizes and on various types of bases, including cloth, paper, extra-thin paper, waterproof paper and cloth, clear film, matte-base film, dimensionally stable Mylar, and acetate. All produce opaque images on translucent or transparent backgrounds. Some are flexible in that undesirable portions of the image can be removed and additions made in ink or pencil, thus making them useful for drafting purposes.

If we were to select the one material considered of the highest quality for producing intermediates by the direct contact printing method, it would have to be acetate contact film. To obtain maximum benefit from this material, however, it must be exposed in a vacuum frame to ensure perfect contact between the negative and film and by a pinpoint light source to ensure preservation of image detail. (See Sec. 3.4, for details of employing pinpoint light sources for exposure.)

In addition to those materials specifically designed for the direct contact printing method, any lithographic film could also be used. This produces film positives. However, lithographic film presents certain disadvantages. For one, it must be developed in a two-part lithographic developer, whereas with regular contact film and other intermediate materials any standard paper developer may be employed. Furthermore, no changes or additions can be made on a film positive. Thus, the image must stand as reproduced from the original.

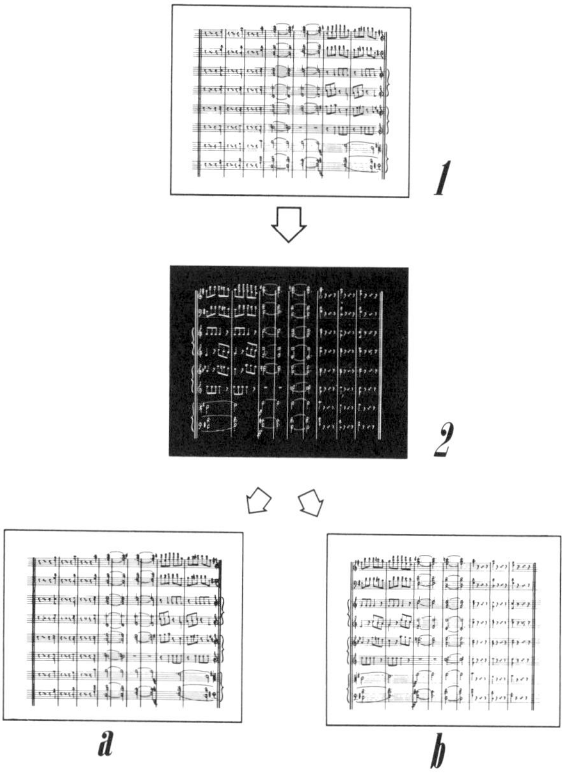

FIG. 8.2 Samples of intermediates made by direct contact printing method. (1) A reverse-reading original; (2) a paper or film negative made from the original; (a) a reverse-reading intermediate; (b) a right-reading intermediate.

Nearly all materials for making intermediates by direct contact are designed for handling under darkroom conditions. Generally, however, the level of illumination can be higher than that required for standard photographic papers. Some materials must be handled under a red safelight, others under a yellow light, and still others under low-level white light. The instructions accompanying each type of material should be checked to determine the type of safelight to employ with that material.

The exposure light source need not be elaborate or exacting, except when using acetate contact film as the intermediate material. Contact materials are not as sensitive as photographic paper or film. Use a lamp with sufficient output to give reasonable exposure time for these slow-speed materials.

If a contact printer is used for exposure, it probably possesses a built-in light source and presents no problem. When using a vacuum frame, however, an external light source is usually required. The light can be mounted on the ceiling or wall, but it should be aimed at the center of the frame. Figure 8.1 is an example of a lighting setup that can be used.

One important fact must be kept in mind concerning exposure: exposing longer than necessary results in overexposure, which produces a less-than-sharp image. It is best, therefore, to always make a test strip before proceeding to the finished product.

The negative being exposed affects overall exposure as much as the sensitivity of the intermediate material. A film negative on an acetate base requires less exposure than paper negatives since the clear acetate image area transmits more light in a shorter time.

There are two ways in which to transfer the image from negative to intermediate material when using the direct contact printing method: right reading and reverse reading (Fig. 8.2).

In producing a right-reading intermediate, the emulsion side of the negative is placed in contact with the emulsion of the material (Fig. 8.3). When the intermediate is fed into the diazo or blueprint machine, it is positioned on top of the sensitive material with the image (or emulsion) facing up and right-reading; that is, so it can be read in a normal manner. The emulsion of the intermediate is not in direct contact with the sensitive diazo or blueprint paper. This method may result in less-than-sharp images since the exposing light of the machine must be transmitted through the intermediate's base and may spread.

To offset this effect, many reproduction engineers prefer the reverse-reading intermediate—one in which the image is backward on the intermediate. In this position, the image must be in direct contact with the sensitive diazo or blueprint paper when run through the machine in order to produce right-reading copies. This direct contact produces a sharper image on the copy.

To make a reverse-reading intermediate, the position of the negative is reversed in the vacuum frame or contact printer, that is, the backside (nonemulsion) of the negative is in contact with the emulsion side of the intermediate material (Fig. 8.3).

In addition to the chemical solutions recommended by the manufacturers on their instruction sheets, practically all intermediate materials used in direct contact printing can be processed in the same chemicals used for photographic paper processing: a standard paper developer, stop bath, and fixer. Although a special developer may be recommended, most instructions will add the note that any MQ-type paper developer—a standard solution—will produce excellent results.

FIG. 8.3 Position of materials to produce right-reading and reverse-reading images by direct contact printing.

DIRECT CONTACT---(Right Reading)

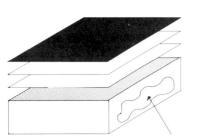

Black Paper Backing
Contact Paper or Film--emulsion down
Negative--emulsion up

Glass Top

Light Source

DIRECT CONTACT---(Reverse Reading)

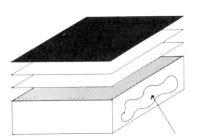

Contact Paper or Film--emulsion down
Negative--emulsion down

Glass Top

Light Source

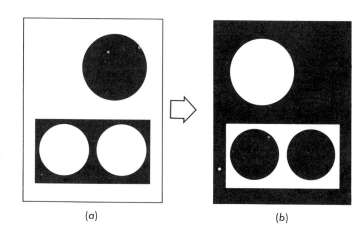

FIG. 8.4   Sample of reflex printing. (a) The original; (b) the negative made from the original by reflex printing.

(a)   (b)

8.2   Reflex Method (Contact Printing)

The reflex method of making an intermediate is usually employed when time is of the essence and quality is not a primary consideration, conditions which lend themselves to the so-called "quick and dirty method" of reproduction. The process can only be employed when a same-size (1:1) print or prints of an original will do.

The result of reflex printing is a reverse-reading paper reproduction with either a reverse clear (or white) image on a black background or a black image on a clear (white) background, depending on the makeup of the original. The reflex reproduction is a reversal of the original (Fig. 8.4). This reproduction can be used in making final prints by the direct contact printing method discussed in Sec. 8.1, or it can be reproduced onto translucent material for use in a diazo or blueprint machine.

The overall quality of the best reflex print is usually not as good as the average reproduction made by other techniques as, for example, camera, autopositive, or photostat. When quality is not the important factor, however, but speed in reproducing an original is essential, reflex printing is a satisfactory method to employ.

The reflex technique is an ideal method for photographic facilities that do little copying and may not have proper camera equipment for reproduction. Photographic studios, for example, frequently use reflex printing when asked to do copying, and many that advertise "photostat services" actually supply simple reflex prints.

Reflex printing is most often used to copy an original that is on opaque stock or one that has printing on both sides. In the former case, the original cannot be contact-printed onto a negative by conventional lighting methods since the opaque stock will not transmit light. In the latter case, the printing on both sides of a page would print through onto the reproduction if regular contact printing were used.

Reflex prints can be produced in either a vacuum frame where the light source is positioned above or in front of the copy or in a reflex printer where the light source is positioned below the copy.

When reproducing opaque originals in a *vacuum frame*, the original is placed into the frame face up. The sensitive reflex paper is then positioned over the original with its emulsion down, that is, with the emulsion in con-

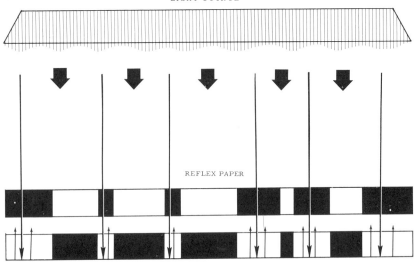

REFLEX PAPER

FIG. 8.5 Position of materials for reflex printing in a vacuum frame.

ORIGINAL

tact with the original's image (Fig. 8.5). The brightest light source available should be used to permit short exposure times. Close the frame, and make the exposure.

If reproducing a double-sided original that is on thin stock in a vacuum frame, follow the procedure explained above. However, a sheet of black paper should be on the frame beneath the original to prevent the printing on the original's reverse side from showing through.

In a *reflex printer,* the procedure is the reverse of a vacuum frame since the light source is from below. Place the reflex paper on top of the glass with its emulsion facing up. Now, position the original on top of the reflex paper with its face down (Fig. 8.6). The exposure is made.

FIG. 8.6 Position of materials for reflex printing in a reflex printer.

REFLEX COPY---Single Side or Opaque Original
(Reverse Reading)

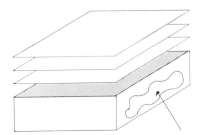

Original--face down
Reflex Paper--emulsion up
Glass Top

Light Source

REFLEX COPY---Double Sided Original
(Reverse Reading)

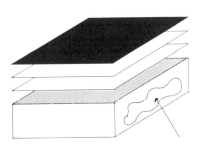

Black Paper Backing
Original--surface down
Reflex Paper--emulsion up
Glass Top

Light Source

The same procedure is followed when reproducing double-sided originals on thin base paper, except that a sheet of black paper is placed over the original to prevent its opposite side from reproducing.

During exposure, the same events occur whether a vacuum frame or reflex printer is used. Light is directed through the base of the reflex paper. The original reflects light back to the reflex paper in those areas where there is no image, that is, from those areas that are clear or white. The image of the original, which is dark or black, does not reflect back light. Consequently, these areas will appear unexposed on the developed reflex print. The result, therefore, is a reverse-reading paper negative.

After exposure, the reflex paper is developed in a standard paper developer. The resulting print possesses a black background area on which is a white (or clear) image area. This paper negative is then used to produce positive prints by contact printing or to make an intermediate on translucent stock by contact printing for use in a diazo or blueprint machine.

It is possible to remove sections of the original image if a translucent intermediate is going to be made. After the paper negative is made, the unwanted area can be removed with opaquing fluid. When the final translucent intermediate is made, changes or additions are then inserted with drafting tools.

If a suitable printer for making reflex prints is not available, these prints can still be made by placing a sheet of black paper on a table top. Place the original, image up, on the black paper, and put the reflex paper over the original with its emulsion down. Lay a sheet of plate glass over the setup to keep it flat. Hold the glass at the edges during exposure to ensure good contact, and place a No. 2 reflector flood bulb about 5 or 6 ft *over* the setup.

If an original printed on both sides is being reproduced onto reflex material using this method, the reverse side should be backed up with black paper. This keeps printing on the reverse side from showing through and exposing on the reflex paper.

8.3 Production of Intermediates by Enlarging

In industry, a need often exists for copies of document data and engineering drawings recorded on reduced-size negatives or positives, particularly microfilm which is used extensively to record great quantities of technical information. In addition, technical data is often recorded on reduced-sized lithographic negatives that may have been used in reports.

Copies of this original data for distribution or file purposes can be made available by production of an intermediate using photographic enlarging techniques (Fig. 8.7). The procedure in the production of intermediates by enlarging is the same as when performing conventional photographic enlarging. The only difference is the material used. If the intermediate is being prepared for use in a diazo or blueprint machine, the material will have to possess a translucent or transparent base.

In projecting microfilm negatives for the production of an intermediate that can be used in a diazo or blueprint machine, a microfilm projector

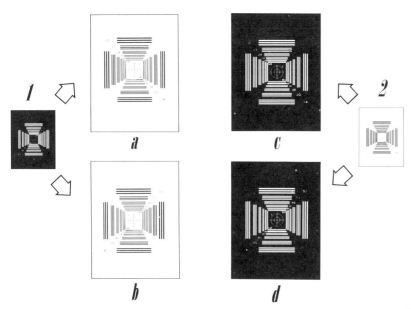

FIG. 8.7   Enlarging techniques. (1) a reduced-size film negative; (a) an enlarged right-reading positive made from the negative; (b) a reverse-reading positive made from the negative; (2) a reduced-size positive from which is made a right-reading (c) or a reverse-reading (d) negative. (a and b) Intermediates for use in a diazo machine; (c and d) intermediates suitable for blueprinting.

(enlarger) is best suited. This equipment is fitted with suitable lenses that produce large images at relatively short projection distances. If such a projector is not available, a regular 35-mm enlarger equipped with a short focal length lens and a roll adapter to hold the microfilm-feed and take-up reels can be used.

As mentioned, projection has to be on a translucent or thin-base material if the reproduction is to be used for making of additional prints in a diazo or blueprint machine. If only one print of the original is needed for reference or file purposes, however, it can be made on any enlarging paper of document weight.

Originals produced on microfilm positive material can be projected to make negative intermediates. These can be used in a blueprint machine to provide positive prints, or they can be contact-printed to provide positive prints or positive intermediates that can be used in a diazo machine. Microfilm positives can be projected onto autopositive materials as well as to make direct positive intermediates (see Sec. 8.6).

In producing intermediates from reduced-size lithographic negatives, a standard enlarger is used instead of the small microfilm enlarger. The materials and techniques used in the projection of microfilm material are the same.

One word of caution: never try to use contact materials when using the enlarging method since exposure times will be too long. There are many translucent-base papers on the market of projection speed that will serve the purpose.

When enlarging, remember that a reverse-reading image will provide maximum printing sharpness. To produce a reverse-reading image, simply "flip" the negative or positive in the enlarger so the projected image is backward.

8.4 Bichromate Hypo Reversal Method

Another method of intermediate production is the bichromate hypo reversal technique, which provides certain advantages not offered by other methods. The process results in a direct positive on orthochromatic lithographic film, which has a clear acetate base. Perhaps the major advantage of this method is its ability to provide high-quality film positives, possessing ultrafine detail, of reduced or enlarged images at a reasonable cost. Since lithographic film can be exposed in the camera, large-size originals can be reduced or small size originals can be enlarged, with the resulting film positive being an excellent intermediate for diazo or blueprint reproduction.

In addition to its usefulness in diazo and blueprint reproduction, the film positives produced by the process can also be used in the preparation of many intermediates used in the graphic arts.

If desired, expose the lithographic film in a vacuum frame. Naturally, no reduction or enlargement can be made, and since the positives are the same size as the original, other techniques such as autopositive explained in Sec. 8.5, may be more practical.

It would be impractical to use the bichromate hypo reversal process for any but large-quantity jobs. If only one or two originals are to be reproduced, it would hardly be inexpensive or a saving in time to mix the various chemicals and go through the necessary steps needed to complete the task.

In addition to a standard lithographic developer and fixing solution, two special solutions are needed. These are a bichromate bleach and a 1 per cent hypo solution, which are mixed as follows (for mixing solutions less than 10 gal, reduce each weight proportionately or measure to the desired proportions):

Bichromate Bleach	Avoirdupois or U.S. Liquid Measure	Metric Measure
Potassium bichromate............	4 oz (powder)	3 grams
Sulfuric acid (concentrate)*......	5 oz (liquid)	4 milligrams
Water to make.................	10 gal	1 liter

* Add sulfuric acid to the solution slowly and with continuous stirring to prevent splashing on parts of the body and clothes.

Hypo Solution	Avoirdupois or U.S. Liquid Measure	Metric Measure
Sodium thiosulfate (anhydrous)...	13 oz (powder)	10 grams
Water to make.................	10 gal	1 liter

Eleven steps are needed to make an intermediate using the bichromate hypo reversal process. These are as follows.

Step 1—Exposure. When exposing the lithographic film in a camera, increase the exposure by 5 per cent from the normal meter reading. If the lithographic film is being exposed in a vacuum frame, place the film into the frame so its emulsion is in contact with the original. Increase exposure 15 per cent more than normal. Exposure is increased to expose as much silver as possible, so that in the bleaching process as much silver as possible is washed away from the background. Only silver that has been fully exposed will be washed away, providing good separation between background and image.

Step 2—Development. Develop the exposed film in a lithographic developer. If exposure was correct, developing time should be completed in about 2½ min with the solution temperature at 68 to 70°F. The film should be agitated frequently.

Step 3—Rinse. Rinse the film in running water at about 68 to 70°F for 1 min. Do not use an acetic acid stop bath to stop development.

Step 4—Bleach Bath. Place the exposed film in the bichromate bleach bath and agitate it for from 15 to 20 sec longer than it takes to clear the image. The bleach bath should be placed in a clean tray that has no bare metal or rust spots. Do not put hands into the bleach unless wearing rubber gloves.

Step 5—Rinse. Rinse the film in running water for 1 min. Once the rinse is started, turn on room lights, and complete the remainder of the process under light. Room lights serve as the source for re-exposing the film's unbleached silver halides. If several films are in process, shuffle them around in the rinse so that all are exposed to light.

Step 6—Fixing. Immerse the film in the 1 per cent hypo bath, and agitate until all background haze from the film has disappeared. As soon as the haze is gone, remove the film from the bath. This part of the process should not be prolonged or some of the unbleached and undeveloped silver particles will be removed, resulting in a loss of density on the final positive.

Step 7—Rinse. Rinse the film in running water for 1 min.

Step 8—Redevelopment. Place the film back in the lithographic developer for 2 or 3 min with agitation. Redevelopment produces a dense black image in the areas that were previously clear on the negative. If faster development is desired, use a universal MQ developer diluted 1 part developer to 2 parts of water.

Step 9—Fixing. Fix the film in a standard hardening fixing bath for lithographic films for about 2 min. If you employ a rapid fixing bath containing ammonium thiosulfate, reduce fixing time to 1 min to guard against bleaching.

Step 10—Washing. Wash the film positive in running water at 68 to 70°F for 15 to 20 min.

Step 11—Drying. Hang the film up to dry in the normal manner. If the film is to be squeegeed for rapid drying; be careful not to damage the emulsion.

8.5   Autopositive Method of Intermediate Production

Perhaps the most widely used process in the production of intermediates is the autopositive method. The term *autopositive* was originally intended as a trade name referring to the Eastman Kodak Company product. Through usage and application, however, the proper noun was transformed into a common one and applied to the process itself.

Several manufacturers make sensitive materials for the process, materials that do not carry the term *autopositive*. For example, Du Pont calls its material Cronoflex direct positive film, while Gevaert refers to its product as Autoreversal. Kodak, of course, still calls its material *Autopositive*.

Autopositive is a valuable reproduction technique, which a technician can use to produce a positive from a positive original without first making a negative, a negative from a negative original without an intermediary step, a negative from a positive, and a positive from a negative.

Most autopositive materials are "color-blind" in that they are sensitive only to blue light. In manufacture, autopositive material is given an overall exposure that produces a latent (invisible) blackening over its surface. If the reproduction technician took a sheet of autopositive material and developed it without first exposing it, the material would develop out black.

If, however, the same sheet of material were first exposed to an original by yellow light, to which it is insensitive, the latent blackening would be destroyed in those areas that receive exposure. When the material is developed, the area that received the exposure would develop out clear, while the area covered by the original's image would develop out black, since it received no exposure.

The yellow-light exposure is a straightforward means of using autopositive and produces a positive image from a positive original, or a negative image from a negative original. The principle of autopositive, then, is quite similar to that of diazo in so far as bleaching of the emulsion is concerned (Chap. 7).

After an exposure has been made and before development, areas bleached by yellow light still retain their blue-color sensitivity. These areas can still react to blue and ultraviolet light or to the blue rays of white light. In this state, the material assumes the same characteristics as any conventional photographic material, with a negative original producing a positive autopositive print or a positive original producing a negative autopositive print.

The basic exposure principles of autopositive material can be summed up as follows: the latent emulsion is destroyed by yellow light, but a reblackening of the emulsion is produced by re-exposure to blue and ultraviolet light or to the blue rays of white light. The effects of yellow and blue light ray exposures upon autopositive material, and their use in combination with one another, are seen in Fig. 8.8.

To take full advantage of autopositive material one should have a means of providing both yellow and blue light ray exposures. Several contact printers manufactured specifically for use with autopositive materials provide both yellow- and white-light sources (which emit blue rays) and can also be used for reflex printing (Sec. 8.2).

A vacuum frame, which provides excellent contact between original and autopositive material, employing a bank of R-2 reflector flood lamps or 500-watt 3200°K lamps in reflectors, can also be used (Fig. 8.9). Lamps should be positioned about 15 to 20 in. apart, center to center, and the overall size of the bank should be about the size of the largest origi-

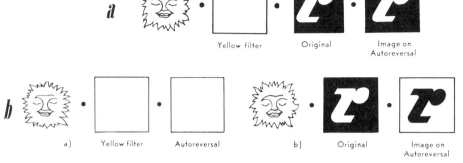

a

Yellow filter · Original · Image on Autoreversal

b

a) Yellow filter · Autoreversal   b) Original · Image on Autoreversal

FIG. 8.8 Effects of yellow and blue light ray exposures on autopositive material. Case A shows the normal method of exposing autopositive materials: The original is exposed through a yellow filter, and the yellow rays transmitted by the letter r locally destroy the latent blackening. Case B shows a less common exposure of autopositive: The emulsion is exposed through a yellow filter, thus destroying the latent blackening over its whole surface. Then the blue-sensitive emulsion is exposed through the original (without a filter). Rays transmitted by the letter r locally re-form a latent image. (*Courtesy of Gevaert, Inc.*)

nal being copied or, more practical, the same size as the vacuum frame. When exposing, the bank of lights should be positioned 24 to 36 in. from the frame.

To make a yellow-light exposure using a vacuum frame, employ yellow bulbs, or cover the original with a sheet of yellow acetate that is manufactured specifically for use with autopositive materials (Fig. 8.10). Acetate is available from graphic-art supply dealers.

Autopositive materials can also be exposed in a diazo machine, but only if a yellow acetate sheet is placed over the original. This method is frequently used in reproducing large engineering drawings. Autopositive materials for this method of exposure are available in rolls, as are yellow and orange acetate cover sheets.

Another popular light source for exposure is the carbon-arc lamp. Usually, a single-rod 35-amp lamp is placed about 18 in. from the copy easel to provide reasonable exposure times. Again, a yellow acetate sheet is placed over the original for the yellow-light exposure.

All light sources heretofore mentioned are suitable for both yellow-light and reblackening (blue-ray) exposures, with the exception of diazo. The

FIG. 8.9 Use of vacuum frame in making an autopositive exposure. Note the wooden frame holding a sheet of yellow acetate around the vacuum frame for the yellow-light exposure. (*Courtesy of Lockheed Electronics Company.*)

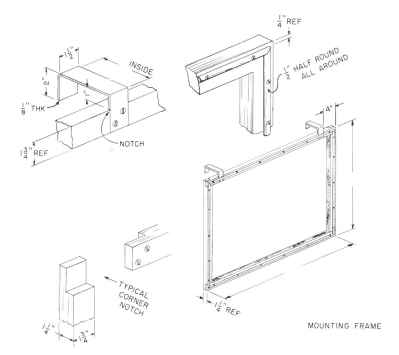

FIG. 8.10   Specifications for fabrication of wooden frame. The frame that holds the yellow acetate sheeting during autopositive exposure in a vacuum frame (Fig. 8.9) can be easily made according to this plan.

reblackening exposure is used primarily for creation of special effects. With other exposure media, aids such as masks can be employed, since the copy setup is in a still position. With diazo, however, it is not possible to use these aids, since the original and autopositive material are in motion. The diazo machine could be used for reblackening, if all one wishes to do is reblacken the entire surface of the autopositive material.

Reblackening of autopositive material (exposure to blue light rays) is accomplished in a shorter time than bleaching (exposure to yellow light). Although it is not possible to recommend exposure times, since certain variables exist from laboratory to laboratory, such as type of original, type of material used, and type of exposure source, certain suggestions to arrive at correct exposure can be offered.

Correct exposure time can be determined most accurately by performing exposure tests. The bracketing method of testing is used, with the first exposure made according to the manufacturer's instructions. Subsequent exposures are made two times, three times, four times, and so forth, greater than the recommended time.

A test strip is made by first covering all but a small section of the material with black paper. After the first section is exposed, the black paper is moved down to reveal more of the material. This procedure is followed until the entire sheet has been exposed.

After the sheet is developed, the best exposure can be determined by examination. If not satisfied with any of the exposures, the test is repeated, using either a shorter or longer base exposure.

The strip method of testing can be employed only when exposures are made in a contact printer or vacuum frame. If a diazo machine is the exposure medium, separate test exposures must be made.

In developing autopositive materials, the manufacturer's instructions should be followed. However, many autopositive materials can be developed in standard M-Q type paper developer, diluted 2 parts of water to 1 of developer, 3 parts of water to 1 part of developer, or 4 parts of water to 1 part of developer, depending upon the material.

Most autopositive material can be developed in a number of developers. For example, Du Pont Cronoflex direct positive film can be developed in a lithographic developer for 1½ to 2 min at 68°F, or in a standard paper developer such as Du Pont 53-D (diluted 2 parts of water to 1 part of developer) for from 30 sec to 1 min at 68°F. Du Pont also recommends development in Du Pont 57-D developer for 30 sec at 68°F.

Kodak autopositive material can be developed in Kodagraph developer (diluted 1:1) for 30 sec at 68°F, or it can be developed in Dektol developer (diluted 3 parts of water to 1 part of developer) for from 30 sec to 1 min at 68°F.

Gevaert autoreversal material can be processed in a lithographic or standard paper developer.

The merits of right-reading and reverse-reading intermediates are discussed in Sec. 8.1. Both reproduction techniques can be used when working with autopositive (Fig. 8.11). To produce a right-reading autopositive intermediate, position the autopositive material, emulsion down, on top of the original, with its face down. To produce a reverse-reading autopositive intermediate, the original's image is placed in contact with the autopositive emulsion. The reverse-reading intermediate provides the sharpest reproduction for use in a diazo machine. In producing reverse-reading autopositive intermediates, it is often necessary to expose longer than for right-reading autopositive intermediates.

Reflex prints of opaque originals or an original printed on both sides can be made using autopositive material. The same procedure outlined in Sec. 8.2 applies. However, a sheet of yellow acetate must be inserted between the light source and the sensitive material if a yellow light source is not provided by the exposure medium (Fig. 8.11).

Many technicians use the autopositive method to improve old, faded, or soiled originals (or another intermediate) when time does not permit the making of a negative. In producing a direct positive intermediate using the autopositive method, two objectives are attained: (1) a translucent intermediate is provided from which other prints can be made; (2) an intermediate of good image density, free of discoloration and other undesirable areas, is provided (Fig. 8.12).

Generally, any autopositive intermediate of a poor original results in a product better in quality than the original, at least in terms of its being free of discoloration. Since direct positive materials are sensitive only to blue-light rays, improved image contrast is provided, since objectionable areas in the background are bleached out by yellow-light exposure. Thus, if one makes a direct positive from a soiled original, most unwanted coloration will be eliminated by means of normal exposure (Figs. 8.13 to 8.15).

Make exposures so that all important information is legibly reproduced. Do not be concerned if some background density is picked up by the intermediate, since it will probably be lost when the intermediate is run through

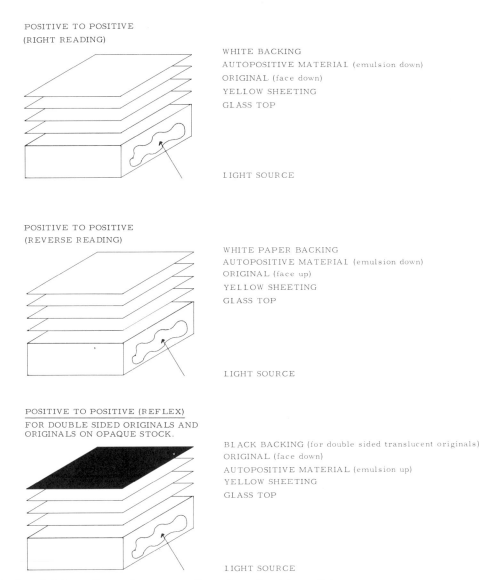

POSITIVE TO POSITIVE
(RIGHT READING)

WHITE BACKING
AUTOPOSITIVE MATERIAL (emulsion down)
ORIGINAL (face down)
YELLOW SHEETING
GLASS TOP

LIGHT SOURCE

POSITIVE TO POSITIVE
(REVERSE READING)

WHITE PAPER BACKING
AUTOPOSITIVE MATERIAL (emulsion down)
ORIGINAL (face up)
YELLOW SHEETING
GLASS TOP

LIGHT SOURCE

POSITIVE TO POSITIVE (REFLEX)
FOR DOUBLE SIDED ORIGINALS AND
ORIGINALS ON OPAQUE STOCK.

BLACK BACKING (for double sided translucent originals)
ORIGINAL (face down)
AUTOPOSITIVE MATERIAL (emulsion up)
YELLOW SHEETING
GLASS TOP

LIGHT SOURCE

FIG. 8.11 Production of right-reading and reverse-reading autopositive intermediates and of opaque originals.

a diazo machine to make additional prints. Badly soiled areas of the original that show up as objectionable spots on the intermediate can be eliminated on the intermediate with chemical eradicators. On the other hand, if some areas of the intermediate appear too light after exposure, they can usually be drawn in by a draftsman.

Exceptionally poor areas of an intermediate can be altered by various means. Ink or pencil can be used to draw in lines; or areas can be eradicated by chemical eradicators and new lines drawn in; or areas can be cut out of the intermediate with a razor blade and a new drawing of the area stripped in. After this touch-up work is performed, a final intermediate can be made.

A technician employing autopositive materials can be creative because of the material's exposure characteristics; that is, the latent blackening is destroyed by yellow-light exposure and the emulsion can be subsequently reblackened by blue-light rays. Keeping these characteristics in mind, a

▣▣ METHOD 1A ▣▣ METHOD 1B

POSITIVE-TO-POSITIVE · RIGHT READING **POSITIVE-TO-POSITIVE** · WRONG READING

Equipment: Vacuum printing frame and 95-ampere carbon arcs, or Diazo printer

Materials: KODAGRAPH AUTOPOSITIVE Film or Paper KODAGRAPH Sheeting, Yellow

PROCEDURE:

Can be performed in room light

1. Place the KODAGRAPH AUTOPOSITIVE sheet, emulsion (shiny side) up, in the printing frame.

2. Place the tracing, face up, over the film or paper.   **2.** Place the tracing, face down, over the film or paper.

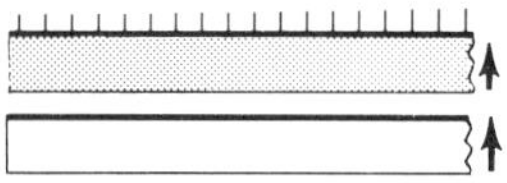

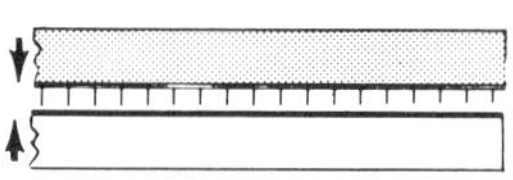

Tracing Face Up

KODAGRAPH AUTOPOSITIVE
Film or Paper, Emulsion Up

Tracing Face Down

KODAGRAPH AUTOPOSITIVE
Film or Paper, Emulsion Up

3. Start the vacuum, and close and swing the frame into position for exposure through KODAGRAPH Sheeting, Yellow.

	Carbon Arcs at 2 feet	Diazo Printer 60 watts per inch
KODAGRAPH AUTOPOSITIVE Paper	35–45 seconds	13 feet per minute
KODAGRAPH AUTOPOSITIVE Film, ESTAR Base	5 minutes	5 feet per minute

	Carbon Arcs at 2 feet	Diazo Printer 60 watts per inch
KODAGRAPH AUTOPOSITIVE Paper (Translucent stock)*	20–30 seconds	15 feet per minute
KODAGRAPH AUTOPOSITIVE Film, ESTAR Base	1½–1¾ minutes	6 feet per minute

*The use of translucent stock offers the further advantage that draftsmen can erase photographic lines with a **moistened** eraser. No chemical eradication is necessary.

4. Process according to instructions packaged with KODAGRAPH material.

CRITIQUE:

Method 1 can be used to best advantage with "Borderline" and "Poor" drawings. It is proper to pick up some background on the AUTOPOSITIVE print, to insure that dimension lines are not lost. This gray, background mottle will drop out when diazo prints are made from the AUTOPOSITIVE intermediate. If a draftsman is to make additions or changes on the AUTOPOSITIVE print, the background can be removed by a ferricyanide solution, such as Farmer's Reducer. The big feature of Method 1 is speed—there is no negative step involved.

FIG. 8.12   Methods of positive-to-positive autopositive reproduction. Procedures can be used in making an intermediate from a poor original. (*Courtesy of Eastman Kodak Company.*)

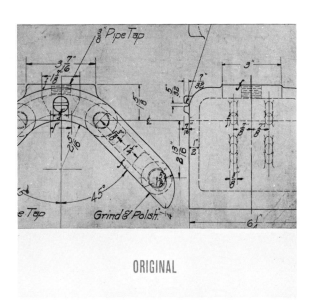

ORIGINAL

FIG. 8.13 Reproducing a poor original. This is a faded original, lacking contrast. (*Courtesy of E. I. Du Pont de Nemours and Company, Inc.*)

"CRONAFLEX"
SECOND ORIGINAL

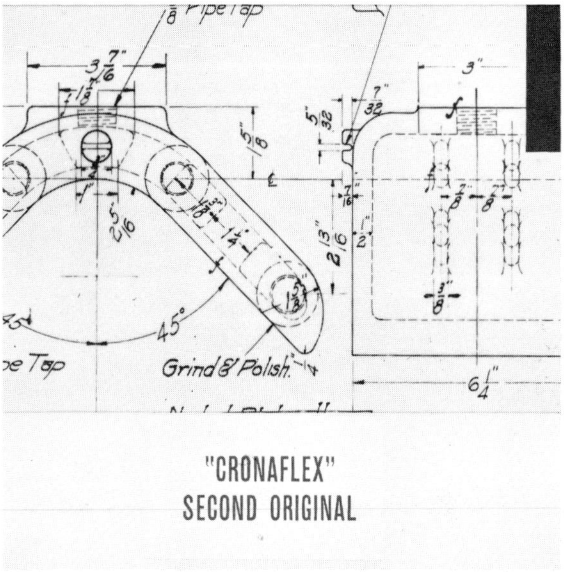

FIG. 8.14 Reproducing a poor original. This is a reproduction of the poor original of Fig. 8.13 made on Cronoflex direct positive film. (*Courtesy of E. I. Du Pont de Nemours and Company, Inc.*)

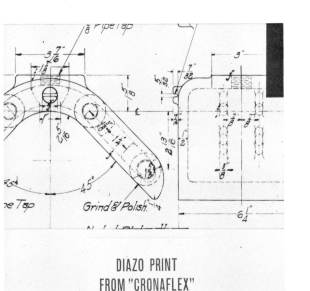

DIAZO PRINT
FROM "CRONAFLEX"

FIG. 8.15 Reproducing a poor original. This is a diazo print of the autopositive intermediate seen in Fig. 8.14. (*Courtesy of E. I. Du Pont de Nemours and Company, Inc.*)

technician can produce positives from positives, negatives from positives, positives from negatives, and combinations. If a mistake is made, he can re-expose the entire sheet of material to blue-light rays, thus reblackening the latent image, and start over. Figure 8.16 illustrates various creative

FIG. 8.16   Various derivations using autopositive material and masking techniques. (*Courtesy of Gevaert, Inc.*)

The negative of the word "geva" was printed through a yellow filter. The yellow filter was then replaced by a negative dot screen. The text negative and the negative screen were finally printed together without a filter.

A line screen was first printed through a yellow filter. The line screen was then removed, and the negative of the word "geva" printed through the yellow filter.

The first steps here are the same as for the illustration above: 1. a line screen and 2. the text negative are printed through a yellow filter. The yellow filter is then removed and replaced by a negative dot screen. Finally, the negative screen and the text negative were printed together without a filter.

The negative of the word "geva" is first printed on Autoreversal through a yellow filter. The yellow filter is then replaced by a negative line screen. Finally, the text and the screen are slightly displaced towards the right and towards the top, and printed without a filter.

The negative of the word "geva" is first copied through a yellow filter. The negative is then replaced by a triangular mask with its sides in register with the lower edge, the righthand edge, and the diagonal of the negative. A further exposure is made through the yellow filter. The yellow filter is then removed and the negative replaced in its previous position. Then the negative and the mask are printed together without a filter.

The text negative is printed together with a line screen through a yellow filter. The line screen is then removed. The text negative is displaced slightly towards the left and towards the top, and printed in this position through the yellow filter.

To obtain the result above, a negative of the word "geva" was printed on Gevaert Autoreversal film, being exposed through a yellow R.488 filter. The yellow filter was then removed, and the text negative re-exposed in such a way that the letter "g" received twice as long an exposure to the light of the lamp as the letters "ev". These in turn received twice as long an exposure as the letter "a". Finally the film was developed in Litholine G 8 developer.

To produce the best conditions for the outline effect, the potassium bromide content of the Litholine developer was increased by adding 25 cm³ of a 10 per cent solution of potassium bromide to 200 cm³ of developer.

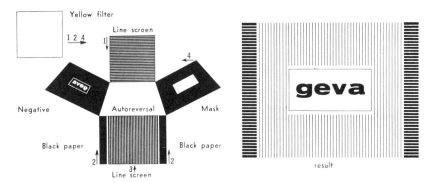

The first exposure produces a negative line screen in the emulsion layer. The second exposure completely destroys this image except underneath the two black strips at the left and at the right. The third exposure produces a positive line screen on the bleached centre strip. The fourth exposure bleaches a small rectangular panel inside this screen. The fifth forms a positive image of the negative lettering within this rectangle.

FIG. 8.17 Examples of masking techniques to produce unusual effects seen in Fig. 8.16. (*Courtesy of Gevaert, Inc.*)

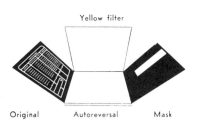

Result

Sequence of steps:

1. Exposure through the negative and yellow filter.
2. Exposure through the mask and yellow filter.
3. Exposure through the negative and mask.

First we have to prepare a mask, so we cut out the area, corresponding to the third column, from a sheet of black paper. The first exposure (through the original and yellow filter) forms a negative image of the original negative in the Autoreversal emulsion. The second exposure is made through the yellow filter and the mask. The aperture in the latter exactly covers the area occupied during the previous exposure by the third column. This second exposure destroys the existing latent blackening over the whole area of the third column. In this area the emulsion again becomes a negative-positive one. The third exposure (through the original and the mask) accordingly forms there a positive image of the third column.

REFERENTIES			
DATUM	ARTIKEL NR.	CODE	BOEK.ST.NR.
-4-3	564.1	32.0	56.31
15-3	432.1	50.4	12.34
18-3	154.6	78.5	34.98
20-3	222.8	35.9	47.23
28-3	354.6	22.5	15.63
-7-4	125.6	32.5	48.96
-9-4	354.6	22.5	15.63
-3-5	125.6	34.5	78.62
21-5	286.5	14.2	74.12
15-6	115.2	12.3	41.23
25-6	123.4	67.2	03.64

FIG. 8.18 Examples of masking techniques to produce unusual effects seen in Fig. 8.16. (*Courtesy of Gevaert, Inc.*)

derivations using autopositive material, while Figs. 8.17 and 8.18 show how to obtain these derivations.

Projection-speed autopositive materials are available that permit the technician to make direct positives by copy camera or photographic enlarger, without a yellow filter. These are considerably more sensitive to light than autopositive materials used in a contact printer or vacuum frame.

Practically all projection-speed autopositive materials require a wash-off stage. After development, the material is sprayed with warm water, which removes the background emulsion, leaving only the image. Development is usually rapid and no fixation is required. For example, Eastman Kodak autopositive projection paper is developed in autopositive projection developer at 71°F for only 60 sec.

Projection-speed autopositive materials permit the technician to make enlarged direct positive reproductions from line negatives or positives and from microfilm. In addition, enlarged or reduced size direct positives for use as intermediates can be made in a copy camera. Since the image is reversed in camera reproduction, it becomes necessary to use an "image reverser" (or prism) over the lens to produce a right-reading positive in the camera. Reproduction of reverse-reading intermediates does not require use of a prism.

Regular and projection-speed autopositive materials are available in a number of bases, including stable-base film, acetate-base film, matte surface film, and extra-thin film, and in heavy and extra-thin weights, and translucent and transparent stocks.

8.6 Photodrawings

Photodrawings are photographs and line drawings, or text, combined on one sheet to supplement each other (Fig. 8.19). The subject is included in this chapter on intermediates since photodrawings are usually made in intermediate form for reproduction by diazo or blueprint. Individual photographs, without text or line drawings, that can be printed by the diazo or blueprint process are also considered as photodrawings by reproduction technicians.

Photodrawings are extremely useful tools to those working in industry, such as draftsmen, engineers, machinists, and other technical types who would have occasion to consult schematics, mechanical drawings, and photographs. Naturally, photodrawings assist photographic departments (industrial, commercial, etc.) burdened by heavy work loads, since they permit an intermediate of a photograph to be made and limited quantities produced by diazo or blueprint machine instead of by means of a time-consuming and expensive photographic printing process.

As indicated, there are several varieties of photodrawings. These can be enumerated as follows:

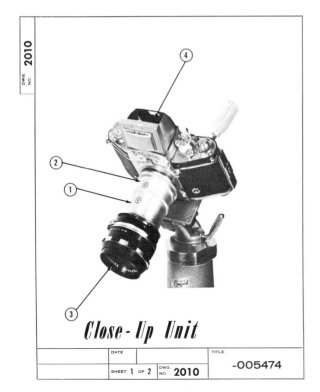

FIG. 8.19   A sample of a photodrawing.

1. Combination photograph and line drawing (or illustration)

2. Combination photograph and text or callouts (without illustration)

3. A photograph reproduced on intermediate material, with space provided on the material for an illustration to be drawn in

Regardless of the type of photodrawing produced, the end result is always the same: an intermediate that will be used for production of additional prints by means of a direct process machine.

One cannot pinpoint all the uses of photodrawings, since use is limited only to the imagination of the reproduction technician. The following, therefore, is an incomplete list, although the more important uses are explained:

1. To aid personnel who cannot "read" or who have trouble "reading" schematics or intricate drawings. By seeing the schematic side by side with a photograph of the item the drawing refers to, individuals can better orient themselves and can obtain clarification of points.

2. To produce reference prints of continuous-tone negatives at low prices when conventional photographs are not required. On these, drafting can be added (or removed) from the intermediate material with no difficulty.

3. To make a perspective drawing of a photograph on the intermediate material on which the photograph is reproduced. This type of combination is often a great aid to those using the photodrawing.

4. To prepare illustrations for low-budget reports and proposals.

Methods of producing photodrawings are many and varied. The method selected depends on the ability of the one doing the reproduction and the equipment available. It is not practical here to explain all methods; therefore the discussion is limited to the most simple techniques, which, incidentally, produce photodrawing intermediates of the best reproduction quality.

Method 1 (Fig. 8.20). The following is a step-by-step procedure to use for this method:

1. Copy a drawing form of the desired size on a sheet of lithographic acetate film.

2. Cut out a portion of the line negative of the drawing form to accept the photographic halftone. Cut out another portion of the line negative of the drawing form to accept the line drawing. (If no line drawing negative is used, the intention being to draw in the line material when the intermediate is complete, just make the cutout for the photographic halftone negative.)

3. Photograph the line drawing, reducing or enlarging it to the required specifications. Make the negative on lithographic film. (Chap. 1 describes reproduction of line material in the camera and the material to use for such reproduction.)

4. Make a halftone negative of the photograph that is to appear on the photodrawing. (Chap. 2 describes the preparation of halftones.) Use either Kodak Autoscreen film or a contact screen, as described in Chap. 2. The halftone image should be reduced (or enlarged) in relation to the size of the line negative. Halftone negatives for photodrawings should have a shadow dot of 30 to 40 per cent and a highlight dot of 80 to 85 per cent

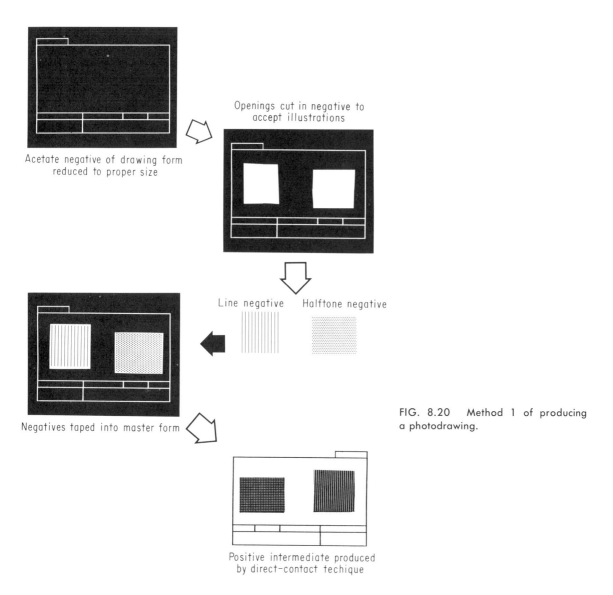

Acetate negative of drawing form
reduced to proper size

Openings cut in negative to
accept illustrations

Line negative   Halftone negative

Negatives taped into master form

Positive intermediate produced
by direct-contact techique

FIG. 8.20   Method 1 of producing
a photodrawing.

(Fig. 8.21). These are the dot sizes most suitable for direct process printing of the photodrawing intermediate by either diazo or blueprint.

5. Tape the negatives into the cutouts.

6. Using the direct contact technique of intermediate preparation (Sec. 8.1), contact print the composite to produce an intermediate. If not included on the negative, the line material required on the intermediate can now be added to the intermediate, or additional drafting, such as specifications, titles, or callouts, can be drawn in.

7. Run the intermediate through a diazo or blueprint machine to produce as many copies as required.

Method 2 (Fig. 8.22). Simple photodrawing intermediates can be made directly from a continuous-tone photographic negative, as follows:

1. Using a standard enlarger, insert the continuous-tone negative into the negative carrier in the standard fashion (emulsion down), and focus the image on a sheet of white paper, which is held in a register frame or

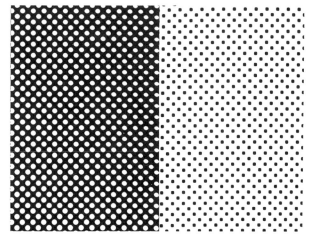

FIG. 8.21   Proper dot structure for photodrawing halftone as it appears on the positive. Left section is 30 to 40 per cent shadow area. Right section represents the 80 to 85 per cent highlight dot.

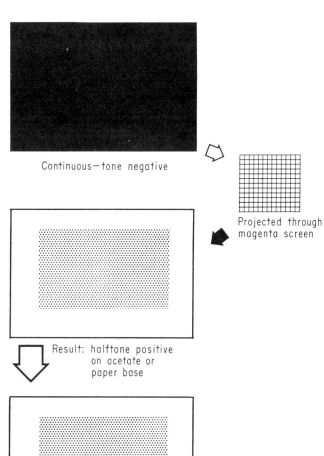

Continuous—tone negative

Projected through magenta screen

FIG. 8.22   Method 2 of producing a photodrawing.

Result: halftone positive on acetate or paper base

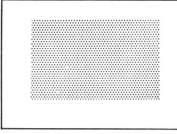

Diazo print

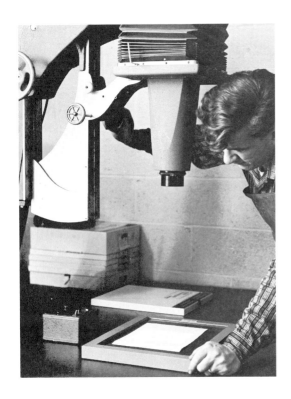

FIG. 8.23 Focusing the image on a sheet of white paper. (*Courtesy of Lockheed Electronics Company.*)

contact printing frame (Fig. 8.23). Adjust the image for size and sharpness, and center it on the white paper. Indicate the position of the register frame or contact printing frame on the enlarger easel by means of pencil marks or tape. This will permit finding of the correct position later on.

2. Place a sheet of projection-speed translucent paper or matte-base lithographic film in the frame (emulsion up) and put a magenta screen on top of it, with its emulsion down (emulsion to emulsion) (Fig. 8.24). Close the frame, so that its glass provides good contact between the screen and sensitive material. In this position, the light from the enlarger's illumination source will pass through the glass, through the magenta screen, and on to the sensitive material.

If translucent paper is being used, employ a coarser contact screen (65- to 100-line) to make the reproduction. This type of screen produces a better image on paper than finer screens. However, if a matte-base lithographic film is used, finer screens may be employed.

3. Place the register or contact frame in its premarked position and make the exposure. Projecting through a magenta screen often requires lengthy exposures. Furthermore, exposure times may vary from negative to negative, depending on negative density, sensitivity of the material, intensity of the light source, and strength of the developer. To determine proper exposure, make a series of test strips.

4. Develop the exposed film or paper in the developer recommended by the sensitive material's manufacturer. The intermediate should possess a dot structure similar in appearance to a halftone negative, but it will be in positive form. Generally, matte-base lithographic films are developed in lithographic developing solutions, which are available in two parts (A and B), in either powder or liquid form. Translucent papers can usually be developed in standard paper developers, diluted as recommended.

5. The intermediate possessing the halftone image is turned over to a

FIG. 8.24 Placing the magenta screen. (*Courtesy of Lockheed Electronics Company.*)

draftsman, who adds the necessary line material, such as titles, callouts, specifications, and so forth, right to the translucent paper or matte-base lithographic film. When he has finished, the photodrawing intermediate is used to make as many prints as necessary by means of diazo or blueprint.

A reproduction technician can use a vacuum printer for reproducing the continuous-tone image on contact-speed transparent paper or matte-base contact film, instead of an enlarger. The magenta contact screen is placed between the light source and the paper, with the negative on top of both; so all material is in the same relative position as when an enlarger is employed. Naturally, in using a vacuum printer one cannot make enlargements or reductions; thus the final product can only be the same size as the negative.

Method 3 (Fig. 8.25). The following method is fairly simple, but effective. It requires less material than methods 1 and 2:

1. Prepare a halftone positive to the necessary size, using the technique explained in method 2. However, the positive should be reproduced on an acetate-base, rather than paper-base material.

2. Obtain a translucent-base drawing form of the desired size. Drawing forms of the type needed are similar in composition to tracing paper.

3. Position the halftone positive on the drawing form, and tape each corner of the positive with small strips of transparent tape.

4. If a drawing is to be included, make a reduced-size negative and contact print it onto the translucent material. Tape this into place on the drawing form.

5. Contact print the entire form onto an autopositive paper or film to make the final photodrawing intermediate, and add any necessary drafting.

Three techniques can be used to make copies of the type of photodrawing intermediate just described. If diazo printing is used, the photodrawing should be kept in positive form.

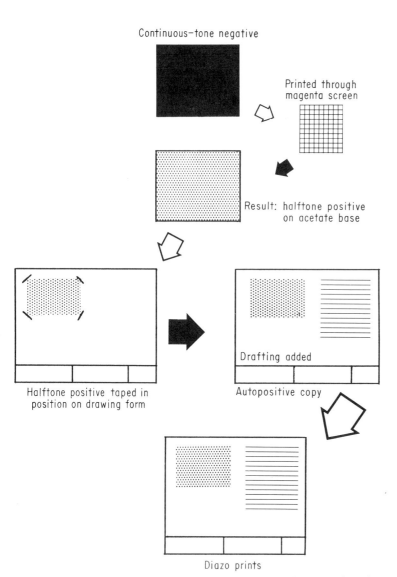

Continuous-tone negative

Printed through
magenta screen

Result: halftone positive
on acetate base

Halftone positive taped in
position on drawing form

Drafting added

Autopositive copy

Diazo prints

FIG. 8.25   Method 3 of producing a photodrawing.

If blueprinting is used as the final printing method, the positive form intermediate should be contact-printed onto a translucent negative stock to produce the final negative. The vandyke method (see Sec. 7.2) is excellent for making reproductions from negative stock. The final negative will produce copies possessing a blue or brown image on a white background when reproduced by blueprint (or vandyke).

If large quantities of the photodrawing are needed, offset printing can be employed. For a positive form intermediate, use a presensitized positive-type offset plate. If the photodrawing was prepared in negative form, use conventional-type presensitized offset plates. It should be remembered, however, that this type of intermediate is not prepared for offset reproduction. Results at best may only be passable. Offset reproduction, therefore, should only be used in emergencies.

appendix

SPECIALIZED
REPRODUCTION PROCESSES

THERE ARE COUNTLESS METHODS and equipment available for reproducing an original. Those discussed in this book are the ones most important to the reproduction technician in industry and government. However, remember that reproduction takes many forms from a typewritten carbon copy to an intricate camera reproduction of a complex line and continuous-tone original.

In recent years, machines to make copying in the office less exhaustive and expensive have been introduced. These are designed to be operated by the average secretary who, in most cases, simply puts an original together with a piece of sensitive material and inserts both into a machine. It is not our intention here to discuss these machines since they are continually being perfected and new ones are being marketed. Besides, they are not considered within the realm of reproduction as this term has been used in this book.

However, there are three important processes which the reproduction technician may find useful in his work, although they are machine processes that one not experienced and trained in reproduction can be taught to employ. These processes are electrostatography (commonly called xerography), photostat, and microfilm.

1. *Electrostatography* is a term used to encompass the entire field that utilizes electrostatically charged patterns to form a visible image. There are many variations of the electrostatographic process. The more popular ones are electrography, electrophotography, and xerography, which is a term credited to the Haloid Xerox Corporation for their significant contributions to this form of reproduction.

The uses of electrostatography can be divided into the following categories: (1) general copying of line originals in which the end product is a duplicate print or prints; (2) preparation of offset masters; (3) enlargements of microfilm; (4) high-speed reproduction of computer information.

How xerography works...

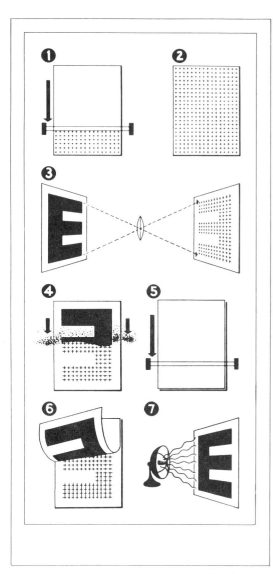

1. Surface of specially coated plate is being electrically charged as it passes under wires.

2. Shows coating of plate charged with positive electricity.

3. Copy (E) is projected through lens in camera. Plus marks show projected image with positive charges. Positive charges disappear in areas exposed to light as shown by white space.

4. A negatively charged powder adheres to positively charged image.

5. After powder treatment (Fig. 4) a sheet of paper is placed over plate and receives positive charge.

6. Positively charged paper attracts powder from plate forming direct positive image.

7. Print is heated for a few seconds to fuse powder and form permanent print.

FIG. A.1   The electrostatographic process. (*Courtesy of Haloid Xerox, Inc.*)

The basic steps of electrostatography can be seen in Fig. A.1, which is aimed specifically at xerography. The process is based on the theory that electric charges of like polarity repel each other, while electric charges of unlike polarity attract each other.

Materials for reproduction by an electrostatographic process are electrically charged. Reproductions are made by manipulating these charges. For example, when employing an electrostatographic camera-reproducer, which is similar in makeup to a graphic-arts reproduction camera (Fig. A.2), the light source acts on the sensitive plate at the film plane to cancel the plate's electric charges. The plate retains its charge in the areas that correspond to the dark sections of the original. This latent, electric image

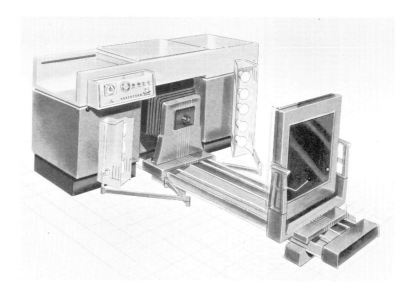

FIG. A.2 An electrostatographic camera-reproducer. All development functions are accomplished in the machine itself. (*Courtesy of Robertson Photo-Mechanix, Inc.*)

permits a charged powder, which is sprinkled on after exposure, to adhere to the plate.

To obtain a copy, a sheet of negatively charged paper is placed over the plate. The powder on the image area of the plate is electrically attracted to the paper and forms an image. The number of copies that can be made with one plate is usually from 8 to 15. Offset plates can be made by this process for larger copy requirements (Fig. A.3).

Reproduction by electrostatography is possible by means of a great variety of specially designed machines, ranging in size from desk-top copiers to large consoles (Fig. A.4). Attachments are also available that permit conversion of a regular graphic-arts reproduction camera to an electrostatographic printer.

Which electrostatographic machine to use depends on the requirements for copies. The important facts to remember in using these machines are as follows:

a. Electrostatography plates are similar to orthochromatic film in sensi-

FIG. A.3 An electrostatographic offset plate maker. (*Courtesy of Haloid Xerox, Inc.*)

FIG. A.4 A typical electrostatographic (xerography) machine for general office use. (*Courtesy of Haloid Xerox, Inc.*)

tivity to the light spectrum. Correction filters can be used with electrostatography plates as they are with orthochromatic films exposed in reproduction cameras. For example, a yellow filter can be used to reproduce blue better.

b. Electrostatography plates exposed in a camera-reproducer should be subjected to white incandescent light as the illumination source. Some office copying units, however, use a greenish fluorescent light source for exposure.

c. Originals that are black on white reproduce best. One can, however, reproduce most types of originals by electrostatography, with the exception of black on red and red on black, since both colors reproduce as black.

d. Electrostatographic printing plates for offset are stable materials and will usually last as long as other types of offset plates.

2. *Photostat* (or stat) is a term accepted through usage and is credited to the Photostat Corporation, which pioneered the process. The photostat machine, which is a photocopying machine, is a highly specialized reproduction camera for producing document reproductions on paper.

Modern photostat-camera equipment is highly sophisticated and employs many automatic features. Most do all processing within the machine itself, meaning that no darkroom is required. Older machines, however, require darkroom processing of paper, since they do not have a self-contained developing chamber.

Photostat cameras employ an optically corrected right-angle prism that reverses an image before it is transmitted to the sensitive paper. The result, therefore, is a right-reading paper negative possessing a white image on a black background.

Sensitive materials used in the process are capable of recording practically every detail found on the original, including various colors, blemishes, and even erasure marks. For this reason, the photostat reproduction is accepted as a legal substitute for official documents, such as birth certificates and service discharge papers.

3. *Microfilm* is a process that depends on a camera which is fitted with an optical system that reduces various size originals to very small size and with extremely high resolution. The first microfilm equipment depended on 35-mm film exclusively. Newer developments, however, make use of 70- and 105-mm film.

Systems in all three sizes are now being used, but for different purposes. Generally, the size one employs depends on the overall requirements. The 35-mm systems are compact, provide small-image sizes that enable reproduction of many frames on a small roll of film, and are less expensive than the larger systems.

On the other hand, the larger-size systems lend themselves best to *blowbacks*, which are enlarged prints. Another use for the larger systems is to provide a reproduction of an original that can be enlarged to the original's size for use as a *second original*.

There are many different types of microfilm systems available in all three sizes, although 35 mm is still the most popular. The latest equipment is highly automated as far as operation and film processing is concerned. Different films are available in all three sizes to provide negative or positive images.

ADDITIONAL WORKS FOR REFERENCE

Chap. 1   Reproduction of Line Material in the Camera

American Technical Society: "A Workbook for the Graphic Arts," American Technical Society, Chicago, 1953.

E. I. du Pont de Nemours and Company, Inc.: "Film Process Control," *Report* A-24770, Wilmington, Del., 1962.

Eastman Kodak Company: "Copying," *Publication* M-1, Rochester, N.Y., 1958.

Graphic Arts Buyer: The Story of Lithography, vol. 3, no. 5, pp. 10–14, 1960, Duralith Publishing Co., Philadelphia.

Karch, R. Randolph: "Graphic Arts Procedures," American Technical Society, Chicago, 1953.

Morgan, Willard D., and Henry M. Lester: "Graphic-Graflex Photography," pp. 177–206, Morgan and Morgan, Inc., New York, 1950.

Wright, George B.: "Copying Technique," chap. 2, American Photographic Book Publishing Co., Inc., New York, 1951.

Chap. 2   Introduction to the Halftone Process

E. I. du Pont de Nemours and Company, Inc.: "Basic Halftone Techniques and the Magenta Screen," *Report* A-13182, Wilmington, Del., 1962.

E. I. du Pont de Nemours and Company, Inc.: *Graphic Arts Pamphlet* A-18999, Wilmington, Del., 1962.

Eastman Kodak Company: *Kodak Publication* 5-56-GLP-E. Rochester, N.Y.

Eastman Kodak Company: *Kodak Publication* P-21 (2-56L-GL). Rochester, N.Y.

Karch, R. Randolph: "Graphic Arts Procedures," pp. 208–209, 218, American Technical Society, Chicago, 1957.

Chap. 3   Camera Reproduction of Printed-circuit Originals

E. I. du Pont de Nemours and Company, Inc.: "Graphic Arts Handbook," section on line procedures of printed circuits. Wilmington, Del.

Eastman Kodak Company: "Photosensitive Resists for Industry," *Publication* P-7, Rochester, N.Y., 1962.

Chap. 4 Silk-screen Stencil Preparation

Eastman Kodak Company: *Kodak Pamphlet* Q-16. Rochester, N.Y.
Eastman Kodak Company: *Kodak Pamphlet* GQ-15. Rochester, N.Y.
Graphic Arts Buyer: Silk Screen Process, vol. 3, no. 5, pp. 20–25, 1960, Duralith Publishing Co., Philadelphia.
Karch, R. Randolph: "Graphic Arts Procedures," p. 8, American Technical Society, Chicago, 1957.
Neblette, C. B.: "Photography—Principles and Practice," pp. 662–672, D. Van Nostrand Company, Inc., Princeton, N.J., 1946.

Chap. 5 Reproduction of Continuous-tone Originals in the Camera

Kramer, Arthur: "Color Photography Techniques," Universal Photo Books, New York, 1958.
McKay, Herbert C.: "Copying," in "The Photographic Negative," vol. 4, pp. 610–612, A. S. Barnes and Co., Inc., New York, 1942.
Shank, W. Bradford: "Filters and Their Use," A. S. Barnes and Co., Inc., New York, 1948.

Chap. 6 Infrared, Ultraviolet, and Radiographs

Eastman Kodak Company: *Kodak Publication* M-3, Rochester, N.Y., 1955.
LaChapelle, Jack: "Solarization," *Report* A-7566, Cowell Memorial Hospital, University of California, Berkeley.
Roderick, J. F., C. W. Walczak, and J. R. Patti: "A Photographic Procedure for Copying Radiographs," E. I. du Pont de Nemours and Company, Inc., *Report* A-6188. Wilmington, Del.,
Shank, W. Bradford: "Filters and Their Use," A. S. Barnes and Co., Inc., New York, 1948.
Tuphaline, C. H. S.: "Photography in Engineering," Faber & Faber, Ltd., London, 1945.

Chap. 7 The Diazo and Blueprint Processes

Coffman, J.: "Technology of the Diazotype Processes," Technifax Corp., Holyoke, Mass., 1961.
Neblette, C. B.: "Photography—Principles and Practice," pp. 149, 354, 357, D. Van Nostrand Company, Inc., Princeton, N.J., 1962.
Ozalid Corporation: Diazo, *Photo Methods for Industry,* March–April, 1961, Gellert-Wolfman Publishing Corporation, New York.
Ozalid Corporation: "They See What They Mean," Ozalid Corp., a division of General Aniline and Film Corp., Johnson City, N.Y., 1959.
Technifax Corporation: "Diazochrome Projectuals for Visual Communication," Technifax Corp., Holyoke, Mass.

Chap. 8 Photographic Methods of Intermediate Production

Eastman Kodak Company: "Photodrawings," *Report* P-22, Rochester, N.Y., 1957.
Eastman Kodak Company: "Kodagraph Reproduction Materials," *Report* 7-52E/JPS-P3, Rochester, N.Y., 1950.

INDEX

70/-

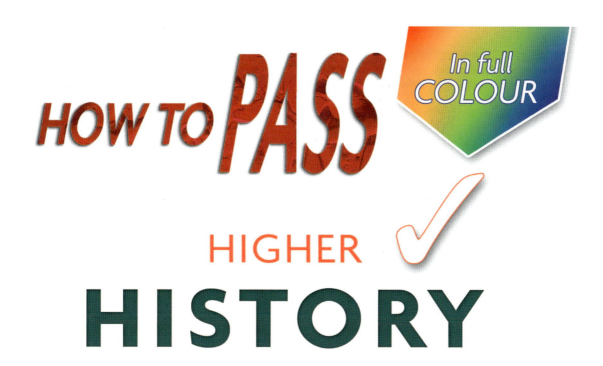

HOW TO PASS

In full COLOUR

HIGHER

HISTORY

Ian Matheson

HODDER
GIBSON
PART OF HACHETTE LIVRE UK

The Publishers would like to thank the following for permission to reproduce copyright material:

Photo credits
Pages 79 and 92 Reproduced with permission of Punch Ltd., www.punch.co.uk; page 81 © Austrian Archives/Corbis; page 82 © Bettmann/Corbis; page 90 NI Syndication.

Acknowledgements
Questions from past exam papers are reproduced with the permission of the Scottish Qualifications Authority.

Every effort has been made to trace all copyright holders, but if any have been inadvertently overlooked the Publishers will be pleased to make the necessary arrangements at the first opportunity.

Although every effort has been made to ensure that website addresses are correct at time of going to press, Hodder Gibson cannot be held responsible for the content of any website mentioned in this book. It is sometimes possible to find a relocated web page by typing in the address of the home page for a website in the URL window of your browser.

Hachette's policy is to use papers that are natural, renewable and recyclable products and made from wood grown in sustainable forests. The logging and manufacturing processes are expected to conform to the environmental regulations of the country of origin.

Orders: please contact Bookpoint Ltd, 130 Milton Park, Abingdon, Oxon OX14 4SB. Telephone: (44) 01235 827720. Fax: (44) 01235 400454. Lines are open 9.00–5.00, Monday to Saturday, with a 24-hour message answering service. Visit our website at www.hoddereducation.co.uk. Hodder Gibson can be contacted direct on: Tel: 0141 848 1609; Fax: 0141 889 6315; email: hoddergibson@hodder.co.uk

© Ian Matheson 2005, 2008
First published in 2005 by
Hodder Gibson, an imprint of Hodder Education,
Part of Hachette Livre UK,
2a Christie Street
Paisley PA1 1NB

This colour edition first published 2008

Impression number   5 4 3 2 1
Year   2012 2011 2010 2009 2008

Cover photo © Photodisc Blue/Getty Images
Cartoons © Moira Munro 2005, 2008
Typeset in 9.5/12.5pt Frutiger Light by Phoenix Photosetting, Chatham, Kent
Printed in Italy

A catalogue record for this title is available from the British Library

ISBN-13: 978-0340-974-032